BASICS OF INSECT MODELING

BASICS OF INSECT MODELING

Edited By

J. L. Goodenough
J. M. McKinion

ASAE Monograph Number 10

Pamela DeVore-Hansen, Technical Editor
Books & Journals

Cover reproduced from a photograph of a southern lubber grasshopper (*Romalea guttata*) taken by James L. Castner, Department of Entomology and Nematology, University of Florida, Gainesville, FL.

FIRST EDITION
Library of Congress Catalog Card Number 92-061514
International Standard Book Number 0-929355-31-8
Manufactured in the United States of America

CONTENTS

CONTRIBUTORS

Allen, J. C., Associate Professor, Department of Entomology and Nematology, University of Florida, Gainesville, FL.

Andrews, G. L., Area Crop Model Specialist, Mississippi Cooperative Extension Service, Stoneville, MS.

Berry, I. L., Professor, Department of Agricultural Engineering, University of Arkansas, Fayetteville, AR.

Brown, L. G., Professor and Head, Department of Industrial Engineering, Mississippi State University, Mississippi State, MS.

Goodenough, J. L., MJ Associates International, P.O. Box 10028, College Station, formerly Agricultural Engineer, Pest Management Research Unit, USDA-Agricultural Research Service (ARS)-SCRL, College Station, TX.

Kline, K. S., Research Affiliate, Department of Agricultural Engineering, Virginia Polytechnic Institute and State University, Blacksburg, VA.

Mack, T. P., Associate Professor, Department of Entomology, Alabama Agricultural Experiment Station, Auburn University, AL.

McCann, I. R., Assistant Professor, Department of Agricultural Engineering, Research and Extension Center, University of Idaho, Moscow, ID.

McFarland, M. J., Professor, Department of Agricultural Engineering, Texas A&M University, College Station, TX.

McKinion, J. M., Electronics Engineer, Crop Simulation Research Unit, USDA-Agricultural Research Service, Mississippi State, MS.

Mishoe, J. W., Assistant Dean (acting) and Professor, Agricultural Engineering Department, Institute of Food and Agricultural Sciences, University of Florida, Gainesville, FL.

Pair, S. D., Research Entomologist, USDA-Agricultural Research Service, Genetics and Production Research Unit, Lane, OK.

Raulston, J. R., Research Entomologist, USDA-Agricultural Research Service, Subtropical Cotton Insects Research Unit, Weslaco, TX.

Smerage, G. H., Associate Professor, Agricultural Engineering Department, University of Florida, Gainesville, FL.

Smith, Jr., J. W., Professor, Department of Entomology, Texas A&M University, College Station, TX.

Sparks, A. N., Research Entomologist (retired), formerly USDA-Agricultural Research Service, Insect Migration/Dispersal Research Unit, Tifton, GA.

Stinner, R. E., Professor, Department of Entomology, North Carolina State University, Raleigh, NC.

Stone, N. D., Associate Professor, Department of Entomology, Virginia Polytechnic Institute & State University, Blacksburg, VA.

Wagner, T. L., Research Entomologist, Crop Simulation Research Unit, USDA-Agricultural Research Service, Mississippi State, MS.

Westbrook, J. K., Meteorologist, USDA-Agricultural Research Service, Crop Insect Pests Management Research Unit, College Station, TX.

Wilkerson, G. G., Assistant Professor, Department of Crop Science, North Carolina State University, Raleigh, NC.

Willers, J. L., Research Entomologist, Crop Simulation Research Unit, USDA-Agricultural Research Service, Mississippi State, MS.

Witz, J. A., Research Scientist, USDA-Agricultural Research Service, Pest Management Research Unit 231, Agricultural Engineering Department, Texas A&M University, College Station, TX.

Wolf, W. W., Agricultural Engineer, USDA-Agricultural Research Service, Crop Insect Pests Management Research Unit, College Station, TX.

Wolfenbarger, Research Entomologist, USDA-Agricultural Research Service, Subtropical Cotton Insects Research Unit, Weslaco, TX.

FOREWORD

Simulation is the inverse of the scientific reductionist process; it is the reassemblage of the individual mechanisms discovered and delineated by reductionist research back into a holistic system. This process of constructing software systems to mimic real-world systems often helps researchers to identify knowledge gaps, for the computer must have equations and numbers. After sufficient knowledge about a specific system has been incorporated into a model, one reaches a "critical mass" of information, after which the model essentially correctly predicts the behavior of the real system. When a model reaches this point is still a matter of interpretation by skillful simulationists, but I suspect some interpretations from information theory could lend scientific and engineering rigor to this area. After one has a working verified and validated model, it is a very useful thing indeed. We have found in working with comprehensive models that one can make discoveries with models. Counter-intuitive processes occurring in the model and in the real world become more obvious because one has the logic of the model (and model output) to trace the event. Finally, we must always remember that modeling is a tool for understanding the performance and interactions of complex systems; it is not an end unto itself.

Many thanks go to all who have contributed to this book. The breadth and depth of their contributions collected here in one place should go a long way to providing those scientists just starting out in modeling and those who want a basic understanding of insect modeling much needed information. I want to express my thanks also to my coeditor, John Goodenough, for inviting me to help edit this book. Finally, thanks go to Martha Barnes and Margaret Lester, my secretary and her assistant, respectively, for their invaluable aid.

James M. McKinion

PART I

STRATEGIES AND METHODS

1

Getting Started: Basics of Modeling Strategies

James M. McKinion

Introduction

A model is a representation of a real system. Models are always less than the real system. How much simplification is allowed is an important topic for each modeling project, but if a simpler model will suffice, simpler is preferred (Phillips et al., 1976). The process of modeling many aspects of insect systems are described in detail in following chapters. The purpose of this chapter is to acquaint you with basic terms and concepts needed for modeling insect systems.

During recent years there has been a large increase in the application of the methods of systems analysis and mathematical modeling for biological and agricultural systems.

A system is a collection of a number of elements or components which are interconnected to form the whole. The behavior of the system not only depends upon the behavior of the individual elements but also upon the nature and the manner of their interconnection. For example, the simulation of agricultural systems, such as crops, insect populations, harvesting systems, etc., involves not only the simulation of individual components of the system but also how these parts relate to one another. Thus, the food substrate that a pest species feeds upon is related to that species' performance in terms of reproduction, movement, and damage to the host crop. The elements of the system work together thereby giving augmentation to the behavior of the whole system.

In referring to a system, one has to specify the boundaries of the system. That which resides inside of the boundaries is the system of interest. That which resides outside of the boundaries is known as the environment. This leads to a natural division of systems into those which are closed and those which are open. An open system is one which is effected by its environment. A closed system is one which is isolated from its environment. In practice, all physical systems in general and all biological systems in particular are open systems. Thus, a model representation of a biological system as a closed system is only an approximation to the real situation.

The most important feature of a systems approach in terms of analysis of a biological system is that all factors affecting the system are considered simultaneously. The exercise of assembling all of the known information about the system and formulating this information in terms of a simulation model forces clear and concise thinking about the system and may emphasize the potential importance of elements of the system and their interconnection which, without the modeling process, may have seemed to be of little importance. Untested assumptions are often highlighted, which in many instances reveals the need for improved experimental design and additional experiments to further delineate the system.

The importance of testing the validity of models must be emphasized. While a model can never be completely validated due to the possibility of non-uniqueness of the model, validation testing enables improvement of a usable model's rejection of obviously false or incomplete models. After a level of confidence in a model has been established through validation testing, then the model can be used not only in a descriptive or explanatory manner, but also can be used to predict the behavior of the system in quantitative terms.

A last important aspect of the simulation of dynamic systems is the concept of feedback. Feedback, one of the most important phenomena occurring in dynamic biological systems, enables the appropriate regulatory functions to be maintained. This can be illustrated by the example of stomatal regulation of transpiration in higher plants. While a plant has a sufficient supply of water and nutrients, during bright sunlight the stomata of the leaves are essentially wide open to allow the maximum uptake of carbon dioxide from the atmosphere for photosynthesis and at the same time allowing the transpiration of water vapor from the leaves to the atmosphere. However, if the plant experiences water stress, the stomata begin to close to help the plant preserve turgor in response to the shortage condition. In this simplified case, leaf turgor is the feedback mechanism.

The appropriate use of mathematical models offers many potential benefits. They provide a concise description of complex processes, indicate ways in which experimental design can be improved, and enable hypothesis testing. Models come in many and varied forms. In the reminder of this chapter, more detailed aspects of modeling philosophy and the different types of modeling strategies are presented.

Simulation Modeling Philosophy

Simulation as used here is a numerical technique for conducting experiments on a digital computer which involves certain types of mathematical and logical models that describe the behavior of business, economic, social, biological, physical, or chemical systems (or some component thereof) over periods of time (Maisel and Gnugnoli, 1972). A simulation model is a particular kind of model of a real system. The model builder constructs the model using a computer language and examines model performance by running the computer program on a digital computer. The major advantage of a simulation is that it permits study of the real system without actual modification of that system in any way. In dealing with a real system, how can one know which changes will result in the desired modified performance of the system? One can use past experience; one can call upon expert opinion (either from a real expert or an expert system implemented upon a computer); one can make changes on a limit if the system is simulated on a computer, the results of the various modifications can be observed in the simulation without modifying the system in any way. Alternative modifications can be tried and their consequences studied in a systematic and controlled manner. Of course, other kinds of models can be used to make predictions about the behavior of real systems. But for complex systems, the use of simulation methodology on a digital computer is unequalled in its ability to provide realistic models of system behavior at a reasonable investment of time and money.

Simulation has other advantages. As a process is studied in preparation for a simulation, previously unrecognized relationships of deficiencies often are revealed. These discoveries may lead to immediate alterations and improvements in the process. Simulations are useful as training devices in teaching about the dynamics of a particular system. Simulation allows the synthesis of the knowledge of scientists from different disciplines into a wholistic description of the system, providing a common communication medium for the different disciplines. Finally, simulation permits the study of a broad range of problems and the asking of extremely complex questions. The computer can manipulate elaborate descriptive and mathematical models that consider a great number of factors, provide for complex relationships, and deal simultaneously with a large number of individual units.

Of course, the powerful advantages of computer simulations are offset by certain disadvantages. These include dollar cost, the use of scarce resources, and the possible long wait before an operational model becomes available. Before the advent of the supermicrocomputer, one of the primary disadvantages was that modeling required the resources of mainframe computers and the associated high costs thereof. However, the continuing progress of computer technology has greatly reduced that cost factor. The expected trend will be that the simulation of even more computationally intensive systems that today require supercomputers will be readily available in the near future (see figure 1). An inspection of technology trends over the last 30 years has shown that one can expect with von Neuman computers (single data, single instruction execution) that every two years a doubling of performance at the same level of cost can be expected (McKinion, 1982). The trend shown in figure 1.1 reflects this statement. However, an entirely new technology situation has arisen which promises to shatter the doubling-every-two years scenario. This new technology is parallel computing developed with the powerful 32-bit microcomputer chips which are now driving the latest generation of personal workstation microcomputers but implemented in a radically new way. These machines are being constructed with large numbers of 32-bit computer chips, each with its own local memory of typically one megabyte, into a massively parallel architecture. As these machines make their way into the market place, greatly

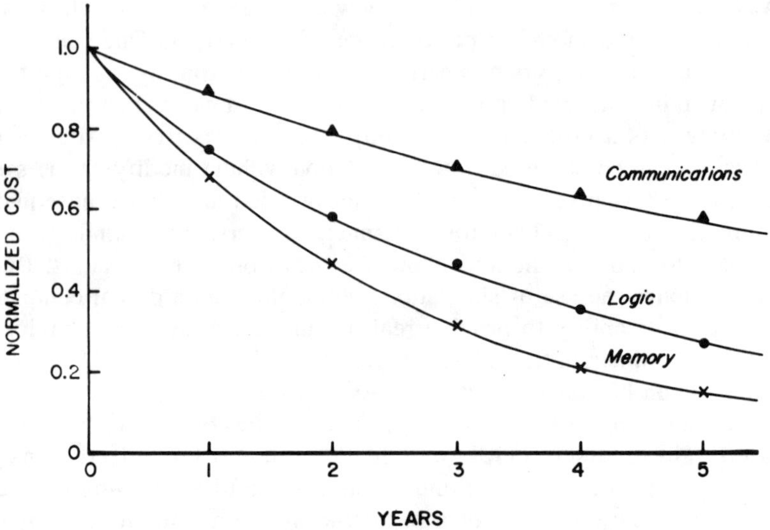

Figure 1.1–Rate of decline of computer costs over a five-year period. Communications reflect the decline in the cost of connecting computers over telecommunications networks. Logic is the cost of the central processing unit chip, and memory is the cost of random access memory chips. The cost is normalized to 1.0 at time t = 0, and the trend lines show the decline one can expect over a five-year period.

increased computer capabilities will be realized and will be available at reasonable costs.

The cost of developing simulations still entails the use of high-priced personnel and extensive studies of operational elements of the system. Because specialists are required to understand and research aspects of the system, this cost probably cannot be mitigated. However, over the years new tools have been developed and are being developed which can help reduce the time element required for system development, testing, and maintenance (McKinion, 1987). The most promising software technology comes from the field of artificial intelligence and expert systems techniques. The object oriented programming paradigm also promise an increasingly important programming methodology as larger and larger systems are simulated which require very large programs (in excess of 100,000 lines of code). Object oriented programming systems (OOPS) have the characteristics of combining procedural code with the data pertinent to that code. Such a device is called an object. When an object receives a message, the procedural code executes.

A more subtle disadvantage of simulation is that the model exists as only a series of computer programs. As a result, management cannot "see" how the simulation actually operates. A result of this is that unless management insists that a model is extensively validated, they cannot tell a good model from a bad

Table 1.1. Summary of advantages and disadvantages of computer simulations vs. experiments with real-world systems

Advantages	Disadvantages
Permits control led experimentation with	Very costly
(a) Consideration of many factors	Uses scarce and expensive resources
(b) Manipulation of many individual units	
(c) Ability to consider alternative policies	Requires fast, high capacity computers
(d) Little or no disturbance of actual system	Takes a long time to develop
Effective training tool	May hide critical assumptions
Provides operational insight	May require extensive field validation
May dispel operational myths	

model or a superior model from an inferior model. It is also easy to overlook the assumptions that were made either explicitly or more subtlely implicitly during model development and the compromises that had to be made to get the model operating on the computer.

A summary of these advantages and disadvantages is shown in table 1.1.

Modeling Paradigms

Models can be thought of as either being simulations of discrete or continuous systems depending upon how the model changes state. An example of a discrete system is an insect population which consists of integer numbers of individual insects in various life stages from egg to larvae to pupae to adults. Discrete systems have variables that can only take on particular values from a finite (but possibly very large) set of alternatives. An example of a continuous simulation is the simulation of pressure in a fluid flow system of a network of pipes. As the liquid flows in a pipe system, changes in pipe diameter, changes in pipe direction (90° elbows), and resistance to flow adjacent to the pipe walls cause a continuous change in liquid pressure from the beginning of the pipe network to the end. Continuous systems have variables which take on any real value in a prescribed interval.

A system may be regarded either as deterministic or stochastic, depending upon the causal relationship between input and output. The output of a deterministic system can be completely predicted if the inputs and the initial state of the system are known. The output of a deterministic system is completely repeatable given the same input conditions are used. An example of a deterministic system is an audio amplifier. When the amplification level is set to a gain of ten (this is the initial state of the system) and a sinusoidal signal of 10 mV peak to peak is applied to the input terminal, an amplified sinusoidal signal of 100 mV from peak to peak will be present at the output terminals. However, a stochastic system in a given state may respond to a given input with an output which is included among a range or distribution of outputs. An example of a stochastic system is the rate of emergence of larvae from eggs of a sample of boll weevils at constant temperature.

Validation

The experimental validation of system models consists of experimentally verifying whether a particular model or type of model postulated for the system represents an adequate description of the system. Basically, models cannot be verified. Models are hypotheses, which are tested by subjecting them to crucial experiments designed to falsify them, and they are accepted to the extent to which they are not falsified. Validation is an integral component of the modeling process, being firmly imbedded within all stages rather than being an activity that is carried out once the model has been developed. Validity is a multi-dimensional concept reflecting model purpose, current theories, and experimental test data relating to the particular system of interest, together with other relevant knowledge (Finkelstein and Carson, 1985).

The criteria with which a model is evaluated can be classified as internal and external. Internal criteria allow conditions within the model itself to be judged without reference to purpose, theory, or data. These criteria are consistency and algorithmic validity. Consistency in this application means that the model contains no logical, mathematical, or conceptual contradictions. Algorithmic validity means that the algorithm for solution or simulation is appropriate and leads to solutions. External criteria refer to model purpose, theory, and data and include empirical validity and theoretical validity. Empirical validity means that the model supports available data. Theoretical validity means that model must be consistent with currently accepted theories or models.

Future of Software Technology

In the past simulation models have been built with conventional computer languages such as FORTRAN, PASCAL, BASIC, and C and with the specialized simulation languages such as CSMP, SIMULA, SIMSCRIPT II, GASP IV, and GPSS. The conventional languages provide portability, availability, flexibility, and speed of execution as their major advantages in building models. The simulation languages provide speed of development (rapid prototyping), integrated graphics, and modeling support tools (specialized algorithms for modeling such as integration or differentiation) as their primary advantages. The disadvantage of using conventional languages is that the length of time to develop the simulation is longer than when using a simulation language. However, the use of a simulation language often limits the modeler's flexibility and requires more computer resources (memory and execution time). A general rule of thumb for using simulation languages is that when doing exploratory modeling or proof of concept modeling, using the appropriate simulation language may be better than developing the model in a conventional high level programming language. If there is a commitment to build a simulation model of a system that is going to require scientist-month(s) of development, building the model using a conventional computer language will be more advantageous (especially where the commitment for modeling the system is essentially open-ended) because of savings in computer time for compiling the model and computer time for running the model.

The recent availability of powerful software tools from the artificial intelligence community in the form of expert systems (ES) and object oriented programming systems (OOPS) promise to greatly enhance simulation methodology. The use of ES with simulation models forms a new AI paradigm for integrating tools for solution of varying and complex problems. The new paradigm is called model-based reasoning. Formally, model-based reasoning was defined as a method in which a detailed structural-functional model of a problem domain, separate from the problem-solving knowledge, was the basis for many different kinds of analysis of the domain. The model was specified in term of structured objects, object behaviors, and relations between the objects (Nardi and Simons, 1986). Model-based reasoning is suitable for complex domains involving one problem to be solved. rather than develop task-specific tools in isolation from one another, a central model provides a unifying foundation for the integration of task-specific tools. Thus, the use of rule-based programming coupled with simulation models promises to provide new and powerful capabilities beyond those currently available. An example of this kind of system is GOSSYM/COMAX, a model-based-reasoning system currently under test for cotton crop production management (McKinion et al., 1987).

The use of object oriented programming systems for development of simulation models, similarly, promises to provide more powerful tools for actual model building. Object oriented programming is a new programming methodology in which a model of a system can be viewed as a collection of system components which are treated as objects. These objects have knowledge contained (programmed) within them about their function. Objects are activated by messages from other objects. The other objects do not have to know anything about the knowledge contained within the object to which messages are passed. The user or programmer of OOPS does not have to be concerned about how the messages are passed. The OOPS environment takes care of that. Current projections for OOPS capability forecast maintainable code for very large systems (in excess of 100,000 lines of code), whereas systems this large using conventional structured programming approaches begin to tax the limits of maintainable code. Another benefit is that OOPS programs operate asynchronously as messages are passed which more closely mimic the real systems. An implied benefit also is that OOPS programs could be directly executed on parallel processor computers without having to elaborate complex schemes to properly segment code as must be done using conventional programming.

Conclusions

Simulation modeling of systems has direct tangible benefits to researchers. The system under study is modeled as a whole, forcing the modeler to evaluate the system in quantitative terms. The model of the system may be easier to understand; it may provide insight into the origins and effects of the process; it may facilitate experimental manipulation of the system and be cheaper than conducting physical experiments. After the model is validated, it can be used to make predictions about how the system operates. New computer hardware and software technology have made modeling methodology more accessible and usable than ever before. The speed and power of desktop workstations and new software tools from artificial intelligence promise to make simulation even more usable and pertinent for solving complex problems than ever before.

References

Finkelstein, L. and E. R. Carson. 1985. *Mathematical MODELING of Dynamic Biological Systems.* New York: Research Studies Press.

Maisel, H. and G. Gnugnoli. 1972. *Simulation of Discrete Stochastic Systems.* Chicago, IL: Science Research Associates.

McKinion, J. M. 1982. Modeling, experimentation, verification and validation: Closing the feedback loop. *Transactions of the ASAE* 25(3):647-653.

McKinion, J. M. 1987. MODVEX: A MOdel Development and Validation EXpert. *Transactions of the ASAE* 30(4):1126-1130.

McKinion, J. M., D. N. Baker, J. R. Lambert and F. D. Whisler. 1987. Expert systems for model validation, development, and application in cotton crop management. In *Proceedings of the 7th International Workshop on Expert Systems and Their Applications*, Vol. 1:187-206. Avignon, France, 13-15 May.

Nardi, B. A. and R. K. Simons. 1986. Model-based reasoning and AI problem solving. In *Proceedings of the Workshop on High Level Tools for Knowledge Based Systems,* Columbus, OH, 6-8 Oct.

Phillips, D. T., A. Ravindran, and J. J. Solberg. 1976. *Operations Research: Principles and Practice*. New York: John Wiley & Sons.

2

The Role and Relationship of the Database with Insect Population Models

Terence L. Wagner and Jeffrey L. Willers

Introduction

"Define the problem clearly.
Start the documentation early.
Think first, code later.
Don't be afraid to start over."
(*Programming Proverbs*)

"The risk of spurious elaboration
can only be minimized by modeling systems analytically,
i.e., by building ever more insight into the construct,
to ensure that new and better observations
result in integration rather than in further patching up."
(F. S. Anderson, 1973)

In many descriptions of simulation modeling (Shannon, 1975; Pritsker and Pegden, 1979; Ignizio, 1982), a sequence of descriptive steps is proposed to conceptualize the flow of information and events. The first step involves establishing communication among people, formulating the problem to be solved, and defining the goals to be achieved. The second step involves selecting the model structure, or type of model, that best represents the system. The problem should determine the appropriate model structure, not a familiarity of the model builder(s) with a given type of model. For models run on digital computers, an abstraction of the system into mathematical or logical relationships is required. These relationships represent both the predictable and unpredictable elements of the system. The third step, and the focus of this chapter, is data acquisition. This step includes the processes of determining what information is required, where it can be found, what needs to be done if the information does not exist, and the eventual collation and transcription of information into a usable form. Additional steps in the simulation process include writing computer code, verifying and documenting code, validating and refining the model(s), conducting simulation experiments, analyzing the results of these experiments, and implementing simulation decisions. Clearly, the process of data acquisition is not independent of the other steps. For example, before determining what type of data are needed, the problem must be well defined (the first step). Data are also used to determine parameter estimates (usually done through regression techniques), initialize, validate, and maintain the model as it is used and refined.

In this chapter, we examine the role and relationship of the database to insect population models. We explore the general nature and type of information used in insect population models, expand the discussion from traditional models to newly developed expert systems, and recommend the use of a dynamic systems approach for modeling insect populations. Although the chapter is not all inclusive, an effort is made to present the critical concepts relating data to model formulation. Nevertheless, it is difficult to improve on the concepts provided by the quotes found at the chapter's beginning, especially if one remembers, to paraphrase the proverb: think first and collect data later.

Background

Entomologists are distinguished among life scientists for their remarkable ability to collect data. In fact, it is reasonable to claim that few life scientists accumulate more information than entomologists. This ability to collect large amounts of information is a mixed blessing, for with it goes the responsibility to effectively integrate and interpret data, and quickly transmit meaningful information to those who can use it. Interestingly, these tasks have been complicated by the constraints imposed by integrated pest management (IPM). As noted by Croft et al. (1976), IPM (as opposed to single-tactic approaches) requires much more information about the status of crops, insects, control strategies, and impacts if optimal decisions regarding insect control are to be determined.

Information Management. Entomology is an applied science where the right information in the right hands can literally save life or livelihood. Perhaps for this reason, entomologists have long recognized the challenge to effectively use information for problem solving; but the complexity of entomological problems has complicated the use of information, a fact that many lay people do not understand or appreciate. Insect problems in agriculture are rarely limited to the occurrence of a single species. Instead, they involve a community of interacting plant and animal species. Understanding species and their interactions in this context is difficult enough, but applying information to management is even harder because the methods for this activity are not well defined. Thus, the ability to manage insect populations below changing (dynamic) thresholds has met with limited success.

Land managers have often dealt with insects in one of two ways: through direct control at times of crisis (e.g., during outbreaks) or through direct control at times when no control is needed (e.g., when populations are endemic or below thresholds). The crisis-oriented manager often deals with low valued crops (such as those found in forestry). This situation suggests a willingness to accept the risk of insect attack. On the other hand, the fear-oriented manager deals with high valued crops (such as those often seen in agriculture) and is often unwilling or unable to accept risk. Unfortunately, both management strategies are inappropriate, but economics are only partly to blame.

A distinction must be made between suppressing insect outbreaks through direct control and preventing outbreaks using a variety of direct and indirect management techniques. Pest and resource managers are engaged too often in controlling outbreaks or applying control when no control is needed. These problems may result from a scarcity of knowledge or an inability to access and effectively use existing knowledge. Thus, the dilemma of having information available and not being able to use it effectively is interesting, but it is unfair to lay blame on the decision makers given the complexities of ecosystems management. The problem simply has had no adequate solution until recently.

Just as synthetic chemical insecticides transformed pest management after WWII, another transformation is presently underway in the area of insect management, brought on by several recent events. The first of these events involves the development of inexpensive yet powerful computers. Computer technology is advancing quickly as computer costs continue to decrease. The second event involves the rapid accumulation of entomological knowledge, some of it in the form of mathematical models developed under various IPM programs of the 1970s (Huffaker, 1980). Finally, and most recently, advancements in artificial intelligence and expert systems have resulted in the marriage of hardware, software, and knowledge-based systems. These events have provided an unprecedented opportunity for pest and resource managers in their quest to apply information to problem solving. The developments have altered the type and form of information available for pest management. The research models developed over the last two decades can now be combined with qualitative information, providing a unique approach to the management of agricultural systems.

Systems Approach to Modeling Insect Populations. A prevailing concept of pest management centers on the regulation of insect populations, which involves an understanding of population dynamics. Population dynamics is the study of change in insect numbers through time and space; therefore, models simulating population behavior must be dynamic. To understand the dynamics of insect populations, we need to consider an insect population as part of a system of interacting components (Getz and Gutierrez, 1982). A system may be defined as a collection of components along with the relationships among components. The description, explanation, and control of system behavior over time is called systems analysis (Curry and Feldman, 1987).

In an early attempt to apply systems analysis to the study of insect population dynamics, Clark et al. (1967) used the term life system. A life system is the part of an ecosystem that determines the existence, abundance, and evolution of a particular insect population. It involves the subject population and its effective environment, representing all the factors that collectively determine the distribution and abundance of a population. In fact, our concept of pest management centers around life systems. For example, as noted by Rabb (1970), pest management is the reduction of pest problems by actions selected after the life system of pests is understood, weighing the ecological and economic consequences of these actions as accurately as possible.

For modeling purposes, system (state) variables are those that are described quantitatively and modeled. Environmental variables drive the system variables. The environment is not modeled; rather, it is monitored. Thus, system behavior is influenced by environment (Curry and Feldman, 1987).

Data Considerations – Developing the Model

An appropriate starting point in modeling a system involves specifying the proposed use of the model, and then defining system boundaries. To a great degree, these steps determine the type(s) of data to be collected. The process of delineating system boundaries (Pritsker and Pegden, 1979) involves understanding the nature of the system; e.g., determining the key system variables to be modeled for a particular purpose, identifying component interactions, and assessing the environmental factors that influence the system variables.

Defining System Boundaries. Once the purpose of the model is defined, we can abstract from complex real-world situations those components and interactions that are essential to the operation of the system, and narrow down the number of variables to those crucial to solving the problem (Shapiro, 1984). A careful review of the real-world system will indicate that some elements are important, and others less important, to satisfying the stated objectives. Nevertheless, the decision of what to include or not include is subjective. According to Shapiro (1984), not enough simplification results in, at best, a costly and overly time-consuming solution; and at worst, one that is too complex and insolvable. However, oversimplification results in the lack of realism and accuracy. Therefore, trade-off decisions must constantly be made in modeling biological systems where realism conflicts with tractability (Curry and Feldman, 1987).

It is important to remember that people, not models, make decisions (Ignizio, 1982), and the value of models in this supportive role depends on the clarity they provide to the decision maker. This clarity is linked to the realism of the model, and an accurate and dependable database represents a tangible connection to the real-world system. (The reader is encouraged to read the "bow tie" example in Shapiro, 1984). Decisions made with good models using unreliable data are just as bad as those made with poor models using trustworthy data. For this reason, one should not become tempted to minimize the importance of data collection or become fatigued from the hard work involved. The exercise of data collection provides insight into system structure that cannot be obtained otherwise.

The variables chosen to represent the system will demarcate what information constitutes the database. For example, if the purpose of a model is to describe the rate of development of immature insect life stages as a function of temperature, then data are required on the development time (or rate) of individuals at different temperatures. Data on mortality rate and fecundity of the adult life stage are not required, and attempts to include this information would encumber the model. On the other hand, if the purpose of the model is to describe the temporal changes in population density, then system variables pertaining to development rate, mortality rate, and fecundity are required for the database. Therefore, system boundaries (determined by the modeler) are closely associated with the content and characteristics of the database.

The foregoing discussion may be oversimplified, but the implications of the concept are important. As an example, Bishop (1972, 1973) describes the results of a model that was used to analyze a cline in industrial melanism in the moth, *Biston betularia.* This deterministic model does not describe the observed cline frequencies of the *carbonaria* form of the moth with increasing distance from Liverpool, England. It could be supposed that this limitation is due to either the omission of an important system variable, or the parameter estimates for some system variables are erroneous. Other examples can be cited which reflect the fact that neither the model nor the database are static entities, but both are dynamic quantities constantly under scrutiny and revision.

A systems approach to modeling insect population dynamics is useful in identifying the important variables that influence the distribution and abundance of insect numbers. Basically, there are three fundamental population processes involved: natality (birth rate), mortality (death rate), and dispersal. These processes can be combined to form a basic demographic equation (a Leslie matrix is one example – Poole, 1974), where population size increases through reproduction and dispersal into an area (immigration), and decreases in size through mortality and dispersal out of an area (emigration). These key processes are influenced by a host of other biotic and abiotic factors, some of which are common to all insect species; others are unique to a given species. A detailed conceptual model of the life history of a species, including insect/insect and insect/host relationships, will help define system components and interactions, and clarify the present state of knowledge. Some variables will be important to the modeling objectives, and a decision must be made as to which will serve as system variables (to be modeled) and which will be environmental variables (to be monitored). In biological systems, multiple environmental variables may influence a single system variable. System or environmental variables may also influence each other, greatly complicating their study and impact on system behavior. Several examples of the processes important to insect population models are presented by Ruesink (1976) and Watt (1961).

Since population models are broken down into components, with each component comprising one or more system variables, the database can be quite variable and reflect information obtained from a broad range of subjects or sources. In other words, the database for a particular model could be comprised of many non-overlapping sets of information (see Watt, 1961, for examples).

Again, the model structure, as determined by the problem statement or goal, should determine the traits of these associated subsets of data.

Another important aspect of system definition is that of the time/space domain of the model. Due to the dynamic behavior of insect populations, the issues of temporal and spatial domain must be addressed early if the proper data are to be collected. Specifying the temporal/spatial framework is important for several reasons: it determines the approach for analyses, the components of the system, and possibly, the appropriate level of detail needed to represent the system (Curry and Feldman, 1987). These issues encompass a range of short or long time frames (e.g., intragenerational or intergenerational considerations) and narrow or broad geographic areas (e.g., within and among fields or within and among regions). If the goal of the model is to examine long time frames over large geographic areas, then several population processes must be considered that otherwise would not be needed. For example, understanding long-term changes in populations across regions requires that diapause, overwinter survival, and dispersal (migration) be modeled. On the other hand, if the goal is restricted to understanding within-field changes in a population over a single growing season, then there may be little need to model the above processes. Instead, these processes may be treated as environmental components that are measured at the appropriate times; e.g., at the start of the growing season to initialize the model. In general, the accuracy and precision of population models increase as the time/space frames decrease.

Models used in IPM often are applied across geographical regions (Welch, 1984). Using cotton as an illustration, many agronomic and entomological differences exist among regions of the cotton belt in the United States. The varieties grown, types of insect pests present, and agronomic practices used are not uniform. These differences alter the model and modify the characteristics of the database used at a single location. For example, the pink bollworm does not occur over the eastern cotton belt. Hence, the analyst should not structure into the model a description of pink bollworm damage to cotton yield in this region.

Information Sources. Many questions should be asked during the process of model development. For example, is the required information already available and in a form usable for the intended purposes? Often, historical records or published results of research can be gleaned for information, with these data being used to initiate the modeling process. It was noted by Innis (1979), however, that modeling begins with a set of objectives and ends with data collection (most of the time), and that data should be collected with particular attention paid to its application. Therefore, as skill with modeling techniques increases, there should be less use of published data that was collected with other objectives in mind.

More often, however, researchers are likely to find that the information needed to develop models simply does not exist, requiring the design and implementation of experiments. If this step is needed, care should be taken to ensure that reliable data are obtained (Bicking, 1954; Bartee, 1968). Data collected by experimentation are frequently the most expensive and laborious to obtain (Pritsker and Pegden, 1979; Berry et al., 1987) and can consume most of the time and resources devoted to solving the problem (Ignizio, 1982). The cost of obtaining additional information is constrained by its usefulness, since the added precision or effort required may not justify the expense (Poole, 1974). Therefore, it is important to consider what information is required to solve the problem. Model building is an iterative process and, as Anderson (1973) pointed out, the improvement of a model needs to be performed thoughtfully. The insight gained by the development of the first approximate model, and by the process of refinement, will result in changes in the database as new experiments are conducted and new information is obtained.

Current Research Trends. An important criterion for interdisciplinary cooperation is effective communication, and the use of a universal medium facilitates communication. Aside from language, the description of energy flow through a system can provide a common meeting ground for the various disciplines involved with modeling insect systems. Analogous to common currency in capital allocation theory (Fleischer, 1984), the calorie may be the best unit to describe energy flow (Paine, 1971). The field of bioenergetics is large and growing (Wiegert, 1976; Calow, 1977; Downer, 1981; Wiegert and Peterson, 1983), and data obtained from investigations on the transformation of energy into insect biomass, maintenance, and reproduction represent tremendous opportunities to the modeler. Just as Sharpe and DeMichele (1977) developed a biophysical model of poikilotherm development, bioenergetics could be used to describe the physiological processes responsible for reproduction, dispersal, diapause, and some causes of mortality in insects. Data from these studies can be extremely helpful in developing realistic models of population dynamics. Representative papers are presented by Gutierrez et al. (1981, 1984, 1987), Reichenbach and Stairs (1985), Waldbauer et al. (1985), Mack and Backman (1986), Slansky (1982), and Willers et al. (1987). Several texts also provide discussions on energetics (Petrusewicz and Macfayden, 1970; Poole, 1974; Southwood, 1978).

Another advancing area of entomological research with application to population modeling involves determining the age structure of adult insect populations in the field (Tyndale-Biscoe, 1984). This determination applies specific traits as estimators of adult age, such as pteridine accumulation (Lehane and Mail, 1985), daily cuticular growth layers (Neville, 1983), or ovarial condition (Grodowitz and Brewer, 1987). Progress in this area should continue and represents information that is of immense value to the system analyst. It must be remembered, however, that most estimates of age represent predictions obtained by "solving the regression equation of Y on X, for X" (Steel and Torrie, 1980). In other words, a value of the dependent variable (e.g., the trait) is used to predict a value of the independent variable (e.g., age). If this procedure is used, the underlying assumptions should be clearly understood, especially if a confidence interval on age is to be determined (Eisenhart, 1939).

Many of the population phenomena of interest require careful planning and long-term commitments from the population ecologist. Research studies can span several insect generations or seasons. The use of well-planned sampling programs (Poole, 1974; Southwood, 1978; Ruesink, 1980) and the collation of life-table data for several generations and localities (Deevy, 1947; Ives, 1964; Harcourt, 1970; Southwood, 1978) represent valuable sets of information to the systems analyst. Such experimental programs challenge the resourcefulness and ingenuity of the investigator, especially given the time, labor, and cost constraints. One representative example is the survivorship study for *Heliothis* larvae on cotton by Hogg and Nordheim (1983). In other instances, useful data on migration, overwintering survival, and the effects of bioclimatic factors and beneficial insects on mortality of pest insects are still lacking (Fye, 1979). Such data on feral adults are inadequate or lacking in many population models, and this deficiency offers excellent opportunity for additional research.

Data Considerations – Operating the Model

Once the form of the model has been determined and its parameter estimates obtained, data sets must be available to initiate and drive the model. Initializing is accomplished by specifying start-up values for system variables. For example, a model that predicts development time of an insect population requires at least two types of data: the starting date of the simulation and a time/temperature file over the range of days of the simulation period. The starting date is used to initialize the model, whereas the time/temperature data are used to drive the model (time and temperature are the independent variables of insect development models). If development times of individual insects are considered, then it is reasonable to use a distribution of oviposition times to apportion the population into cohorts. Multiple starting times are then available to initialize the model (Wagner et al., 1985). In this instance, a single starting time is insufficient, although the time/temperature data used to drive the model remain the same.

For population models that describe numerous processes, the output of one submodel (e.g., oviposition) may be used to initialize or drive another submodel (e.g., development). Generally, as models become more complicated (as they describe the effects of numerous independent variables on a single process or describe numerous processes), more information is needed to initiate and drive them. In fact, the amount of information needed to initiate large models may be cumbersome to users. Therefore, default values or standardized data files are often supplied to simplify the use of models.

In an article on environmental monitoring for pest management systems, Haynes et al. (1973) stated that one of the important factors limiting the usefulness of models in pest management is the ability to obtain real-time information on pest/crop systems. Time is the key, since scientists must work with the climatic and biological data unfolding during the periods when various management strategies can still be implemented. Information that is gathered in the field must be made available quickly to the models. Welch (1984) provides an excellent review on acquiring, processing, and accessing information (data) for use in IPM systems.

Weather Data. Up-to-date weather information is critical to pest management systems in general and insect population models in particular. To one degree or another, most of the principal components of the system (including crop, pests, natural enemies, and control practices) rely on this kind of information to drive system variables. Ambient air and soil temperatures, rainfall, solar radiation, and wind are the most important input variables to models; and of these, temperature is most important to insect models. Often, records from the National Oceanographic and Atmospheric Administration (NOAA) are used as the source of information (Croft et al., 1976), but official weather stations are widely spaced and access to the information is relatively slow. Consequently, the information is available on no less than a daily basis. To increase the time resolution, daily maximum and minimum temperature values are often converted to shorter time steps using mathematical equations (Allen, 1976; Curry et al., 1978; Curry and Feldman, 1987; Taylor, 1981). More recently, some computer-based management systems employ a weather station located on individual farms. Although expensive, the information acquired from these stations is much more pertinent because of its proximity to the area being modeled. Furthermore, data from these stations can be uploaded directly and quickly into computers (Welch, 1984).

Biological Data. Field observations and measurements of crop and insect dynamics are also vital to pest management systems. This information is used for a variety of purposes including initializing and driving models. Data on crop growth and development, crop damage, and pest and beneficial phenology and density are particularly important. For example, pheromone trap captures in early season are sometimes used as a reference of adult phenology to initialize models. Other models (such as those describing reproduction) may require more specific information on adult density and age structure for their use. Because pest population models often perform poorly over the long-term, it is necessary to update the model with current information from the field.

Data Considerations –Testing and Refining the Model

Sensitivity analysis is an important activity in the modeling process. This analysis is performed by varying the values of the input parameters, one or more values at a time, and observing the impact of these changes on model behavior. Broadly speaking, two outcomes are possible: one where small changes in parameter values result in distinct changes in output (hence the model is sensitive to these parameters), or alternately, where extensive changes result in little difference in model output. We agree with Shoemaker (1980) that sensitivity analysis is useful for setting research priorities for obtaining data. The most sensitive parameters of the model should be the most precisely measured.

Another activity that influences the database is the validation of the model (Welch et al., 1981). For clarity, a distinction needs to be made between model verification and validation. Verification refers to the process of assuring that the computer model runs or executes as it should, whereas validation refers to the confirmation that the model is an accurate representation, within defined limits, of the real-world system (Pritsker and Pegden, 1979). Curry and Feldman (1987) examine the statistical problems in validation which have general implications for the data requirements of models.

Models that perform adequately in one geographic region may not perform in another. In this case, adjustments of some parameter values may correct the problem. This procedure is called model calibration. Shoemaker (1980) cautions that the field data used for validation should be different from that used for calibration, and that calibration is a reasonable procedure if the modified values of the parameters do not deviate too largely from their original values. How large is too large is a subjective decision influenced by the conditions under which the parameters were measured (laboratory vs. field) and by the variance of the data. Similarly, the data used to validate the model should be different from the data used to develop the model.

The variance of data can also impact the model parameters in an interesting way. Fye (1979) mentioned that the use of a sequence of regression equations to estimate a final parameter is not without risk. If the variance or confidence limit of one of the regression estimates is broad, an uncertainty is introduced into the final estimate which could result in a departure of the model's prediction from the outcome observed. Another important consideration involves the reduction in the degrees of freedom in the error term, as the number of interactions among the variables in a model increases. Many data sets are required when tests of significance of more than two or three variables are to be performed. [See Mott in Morris (1963), cited by Poole (1974).] One set of data from a single location or timepoint should never be considered sufficient for validation.

Once the system has been implemented, its performance must be monitored and evaluated so that it can be improved. It is important to remember that observations obtained in the laboratory can differ from observations made in the field. Frazer and Gilbert (1976) stated that all laboratory studies must be suspect until verified in the field. Evaluating performance will require feedback from both the system user and the analyst.

Expert Systems

The application of expert systems in entomology is an area of current interest and expanding development (Stone et al., 1986; Coulson et al., 1987). As with simulation models, expert systems require the establishment of a well-defined problem statement, a step that is followed closely by the obtainment of knowledge (e.g., information or data) (Huber and Huber, 1988). Unlike traditional modeling approaches, however, expert systems use information differently. They strive to capture the knowledge, training, and intuition of experts so that non-experts can perform, or solve problems, with similar competence. Typically, this expertise is represented as fact-and-rule definitions which comprise the knowledge base of the system. An algorithm (called an inference engine) controls the selection of information and processes the contents of the knowledge base. Combined, the knowledge base and inference engine distinguish expert systems from other database systems (Huber and Huber, 1988). Depending on how the knowledge is coded, information used by the inference engine to evaluate rules can be imprecise or probabilistic (Stone et al., 1986). Other forms of information comprising the knowledge base can be obtained as output from simulation models, resulting in the models becoming the domain experts. In this case, the accuracy of the database used by the models and the realism that the models provide, play a fundamental role in the usefulness of the system. These integrated systems may provide expanded power and usefulness over rule-based systems or models alone. An example of a model-based reasoning system is GOSSYM/COMAX (McKinion et al., 1989), a decision aid that assists farm managers in various aspects of cotton production. This is one of the few systems of its kind presently in use in agriculture.

Finally, in the struggle to manage information, expert systems can serve as an effective tool for quickly assimilating large volumes of data into meaningful patterns of information, and apply inexact reasoning techniques to propose a likely solution to a problem (Parks, 1987). Given the decision-oriented nature of many entomological problems, the development of expert systems represents a complementary approach for agro-ecosystems management.

Conclusions

As agricultural scientists become more aware of the tools available for modeling systems (DeMichele, 1975; Feldman and Curry, 1982), the processes of data collection, use, and maintenance will become a function of the requirements specified by these tools. In the course of this development, coupling of the basic practices of observation and data collection with model building will result in increased understanding of biological systems (Legendre and Legendre, 1983). The fact that models provide a means for organizing complex systems should result in greater cooperation among the various scientific disciplines involved, providing opportunities for the obtainment and integration of data from diverse sources. In addition, because of the diverse array of modeling approaches, there are also opportunities for artistic endeavors in the development of both the model and its database (Shannon, 1975).

Acknowledgment. We thank D. N. Baker, W. S. Fargo, R. G. Luttrell, G. H. McKibben, and R. L. Olson for helpful reviews of this chapter.

References

Allen, J. C. 1976. A modified sine wave method for calculating degree days. *Environ. Entomol.* 5:388-396.

Anderson, F. S. 1973. Introduction. In *Insects: Studies in Population Management*, ed. P. W. Geier, L. R. Clark, D. J. Anderson and H. H. Nix. 708. Ecol. Soc. Aust. (memoirs 1).

Bartee, E. M. 1968. *Engineering Experimental Design Fundamentals*. Englewood Cliffs, NJ: Prentice-Hall.

Berry, J. S., T. O. Holtzer and J. M. Norman. 1987. Computers in the field. *Bull. Entomol. Soc. Am.* 33:209.

Bicking, C. A. 1954. Some uses of statistics in the planning of experiments. *Ind. Qual. Control* 10:20-24.

Bishop, J. A. 1972. An experimental study of the cline of industrial melanism in *Biston betularia* (L.) (*Lepidoptera*) between urban Liverpool and rural North Wales. *J. Anim. Ecol.* 41:209-243.

———. 1973. The proper study of populations. In *Insects: Studies in Population Management*, ed. P. W. Geier, L. R. Clark, D. J. Anderson and H. H. Nix. *Ecol. Soc. Aust.* (memoirs 1).

Calow, P. 1977. Ecology, evolution and energetics: A study in metabolic adaptation. *Adv. Ecol. Res.* 10:1-62.

Clark, L. R., P. W. Geier, R. D. Hughes and R. F. Morris. 1967. *The Ecology of Insect Populations in Theory and Practice*. London: Methuen.

Coulson, R. N., L. J. Folse and D. K. Loh. 1987. Artificial intelligence and natural resource management. *Sci.* 23:262-267.

Croft, B. A., J. L. Howes and S. M. Welch. 1976. A computer-based, extension pest management delivery system. *Environ. Entomol.* 5:20-34.

Curry, G. L., R. M. Feldman and K. C. Smith. 1978. A stochastic model of a temperature-dependent population. *Theor. Popul. Biol.* 13:197-213.

Curry, G. L. and R. M. Feldman. 1987. *Mathematical Foundations of Population Dynamics*. Mono. Ser. 3. Tex. Eng. Exp. Stn., College Station, TX.

Deevy, E. S., Jr. 1947. Life tables for natural populations of animals. *Quart. Rev. Biol.* 22:283-314.

DeMichele, D. W. 1975. An evaluation of modeling, systems analysis and operations research in defining agricultural research needs and priorities in pest management. *Iowa State J. Res.* 49:597-621.

Downer, R. G. H. 1981. Physiological and environmental considerations in insect bioenergetics. In *Energy Metabolism in Insects*, ed. R. G. H. Downer, 1-17. New York: Plenum Press.

Eisenhart, C. 1939. The interpretation of certain regression methods and their use in biological and industrial research. *Ann. Math. Stat.* 10:162-180.

Feldman, R. M. and G. L. Curry. 1982. Operations research for agricultural pest management. *Oper. Res.* 30:601-617.

Fleischer, G. A. 1984. *Engineering Economy: Capital Allocation Theory*. Monterey, CA: Brooks-Cole.

Frazer, B. D. and N. Gilbert. 1976. Coccinellids and aphids: A quantitative study of the impact of adult ladybirds (*Coleoptera: Coccinellidae*) preying on field populations of pea aphids (*Homoptera: Aphididae*). *J. Entomol. Soc. Brit. Col.* 73:33-56.

Fye, R. E. 1979. Cotton insect populations: Development and impact of predators and other mortality factors. USDA Tech. Bull. 1592.

Getz, W. M. and A. P. Gutierrez. 1982. A perspective on systems analysis in crop production and insect pest management. *Ann. Rev. Entomol.* 27:447-66.

Grodowitz, M. J. and F. D Brewer. 1987. Ovarian anatomy and physiological age-grading of the female boll weevil, *Anthonomus grandis grandis* Boheman (*Coleoptera: Curculionidae*) *Ann. Entomol. Soc. Am.* 80:642-651.

Gutierrez. A. P., J. U. Baumgaertner and K. S. Hagen. 1981 A conceptual model for growth, development, and reproduction in the ladybird beetle, *Hippodamia convergens (Coleoptera: Coccinellidae). Can. Entomol.* 113:21-33.

Gutierrez, A. P., J. U. Baumgaertner and C. G. Summers. 1984. Multitrophic models of predator-prey energetics. *Can. Entomol.* 116:923-932.

Gutierrez, A. P., F. Schulthess, L. T. Wilson, A. M. Villacorta, C. K. Ellis and J. U. Baumgaertner. 1987. Energy acquisition and allocation in plants and insects: A hypothesis for the possible role of hormones in insect feeding patterns. *Can. Entomol.* 119:109-129.

Harcourt, D. G. 1970. Crop life tables as a pest management tool. *Can. Entomol.* 102:950-955.

Haynes, D. L., R. K. Brandenburg and P. D. Fisher 1973. Environmental monitoring network for pest management systems. *Environ. Entomol.* 2:889-899.

Hogg, D. B. and E. V. Nordheim. 1983. Age-specific survivorship analysis of *Heliothis* spp. populations on cotton. *Res. Popul. Ecol.* 25:280-297.

Huber, L. E. and R C. Huber. 1988. Cracking researchers' monopoly on expert systems expertise *Indust. Eng.* 20:58-65.

Huffaker, C. B., ed. 1980. *New Technology of Pest Control*. New York: John Wiley & Sons.

Ignizio, J. P. 1982. *Linear Programming in Single- and Multiple-Objective Systems*. Englewood Cliffs, NJ: Prentice-Hall, Inc.

Innis, G. 1979. A spiral approach to ecosystem simulation, I. In *Systems Analysis of Ecosystems*, ed. G. S. Innis and R. V. O'Neill, 211-326. Fairland, MD: International Co-Operative.
Ives, W. G. J. 1964. Problems encountered in the development of life tables for insects. *Proc. Ent. Soc. Man.* 20:34-44.
Legendre, L. and P. Legendre. 1983. *Numerical Ecology*. Amsterdam: Elsevier.
Lehane, M. J. and T. S. Mail. 1985. Determining the age of adult male and female *Glossina morsitans morsitans* using a new technique. *Ecol. Entomol.* 10:219-224.
Mack, T. P. and C. B. Backman. 1986. Effects of fluctuating diel temperatures on longevity and oviposition rate of adult female lesser cornstalk borers (*Lepidoptera: Pyralidae*). *Environ. Entomol.* 15:715-718.
McKinion, J. M., D. N. Baker, F. D. Whisler and J. R. Lambert. 1989. Application of the GOSSYM/COMAX system to cotton crop management. *Agric. Syst.* (In Press).
Morris, R. F., ed. 1963. The dynamics of epidemic spruce budworm populations. *Mem. Entomol. Soc. Can.* No. 31.
Neville, A. C. 1983. Daily cuticular growth layers and the teneral stage in adult insects: A review. *J. Ins. Physiol.* 29:211-219.
Paine, R. T. 1971. The measurement and application of the calorie to ecological problems. *Annu. Rev. Ecol. System* 2:145-164.
Parks, M. W. 1987. Expert systems: Fill in the missing link in paperless aircraft assembly. *Indust. Eng.* 19:37-45.
Petrusewicz, K. and A. Macfayden. 1970. *Productivity of Terrestrial Animals: Principles and Methods*. IBP Handbook 13. Oxford: Blackwell Scientific Publications.
Poole, R. W. 1974. *An Introduction to Quantitative Ecology*. New York: McGraw Hill.
Pritsker, A. A. B. and C. D. Pegden. 1979. *Introduction to Simulation and SLAM*. New York: John Wiley & Sons.
Rabb, R. L. 1970. Introduction to the conference. In *Concepts of Pest Management*, ed. R. L. Rabb and F. E. Guthrie, 1-5. Raleigh: North Carolina State Univ.
Reichenbach, N. G. and G. R. Stairs. 1985. Bioenergetics of the western spruce budworm (*Lepidoptera: Tortricidae*) with comments on endotherm and ectotherm population energetics. *Can. J. Zool.* 63:1330-1338.
Ruesink, W. G. 1976. Status of the systems approach to pest management. *Annu. Rev. Entomol.* 21:27-44.
———. 1980. Introduction to sampling theory. In *Sampling Methods in Soybean Entomology*, ed. M. Kogan and D. C. Herzog, 61-78. New York: Springer-Verlag.
Shannon. R. E. 1975. *Systems Simulation: The Art and Science*. Englewood Cliffs, NJ: Prentice-Hall.
Shapiro, R. D. 1984. *Optimization Models for Planning and Allocation: Text and Cases in Mathematical Programming*. New York: John Wiley & Sons.
Sharpe, P. J. H. and D. W. DeMichele. 1977. Reaction kinetics of poikilotherm development. *J. Theor. Biol.* 64:649-670.
Shoemaker, C. A. 1980. The role of systems analysis in integrated pest management. In *New Technology of Pest Control*, ed. C. B. Huffaker, 25-50. New York: John Wiley & Sons.
Slansky, F. 1982. Toward a nutritional ecology in insects. In *Proc. 5th Inter. Symp Insect-Plant Relationships,* 253-259. Wageningen.
Southwood, T. R. E. 1978. *Ecological Methods with Particular Reference to the Study of Insect Populations*. New York: Halsted Press, John Wiley & Sons.
Steel, R. G. D. and J. H. Torrie. 1980. *Principles and Procedures of Statistics: A Biometrical Approach*. New York: McGraw-Hill.
Stone, N. D., R. N. Coulson, R. E. Frisbie and D. K. Loh. 1986. Expert systems in entomology: Three approaches to problem solving. *Bull. Entomol. Soc. Am.* 32:161-166.
Taylor, F. 1981. Ecology and evolution of physiological time in insects. *Am. Nat.* 117:1-23.
Tyndale-Biscoe, M. 1984. Age-grading methods in adult insects: A review. *Bull. Ent. Res.* 74:341-377.
Watt, K. E. F. 1961. Mathematical models for use in insect pest control. *Can. Entomol. Suppl. No. 19.* 93:1-62.
Wagner, T. L., H. Wu, R. M. Feldman, P. J. H. Sharpe and R. N. Coulson. 1985. Multiple-cohort approach for simulating development of insect populations under variable temperatures. *Ann. Entomol. Soc. Am.* 78:691-704.

Waldbauer, G. P., J. G. Sternburg and D. H. Janzen. 1985. Longevity and weight loss of free-flying male cecropia moths, *Hyalophora cecropia* (*Lepidoptera: Saturniidae*) *Great Lakes Entomol.* 18:133-138.

Welch, S. M. 1984 Developments in computer-based IPM extension delivery systems. *Annu. Rev. Entomol.* 29:359-81.

Welch, S. M., B. A. Croft and M. F. Michels. 1981. Validation of pest management models. *Environ. Entomol.* 10:425-432.

Wiegert, R. G., ed. 1976. *Ecological Energetics, Benchmark Papers in Ecology*, Vol. 4. Stroudsburg, PA: Dowden, Hutchinson and Ross.

Wiegert, R. G. and C. E. Peterson. 1983. Energy transfer in insects. *Annu. Rev. Entomol.* 28:455-486.

Willers, J. L., J. C. Schneider and S. B. Ramaswamy. 1987. Fecundity, longevity and caloric patterns in female *Heliothis virescens*: Changes with age due to flight and supplemental carbohydrate. *J. Insect Physiol.* 33:803-808.

3

Nonlinear Dynamics and Chaos in Insect Population Models

Jon C. Allen

Introduction

This chapter will take a somewhat theoretical approach in that no particular insect is meant to be modeled nor are the models given sufficient detail to handle the complexity of field situations. What is intended is simply to illustrate the kinds of dynamics that can emerge from simple nonlinear models. Throughout this chapter "nonlinear" will mean nonlinear in system variables not to be confused with parameters being nonlinear functions of environmental variables (e.g., temperature) as is often done in insect models (Stinner et al., 1974; Logan et al., 1976; Sharpe and DeMichele, 1977; Wagner et al., 1984; Allen, 1976, 1988). Forcing of parameters by the environment will be referred to as environmental forcing or periodic forcing, if the functions are idealized to periodic form. The dynamics of a forced system which is linear in the system variables is basically just a stable system that is "pushed around" by the environment (no matter how complicated the forcing). "Stable" is used loosely here to mean decaying to a fixed point (which might be zero). While environmental forcing can interact in highly complicated and surprising ways with a nonlinear system, the interaction with a linear system is basically simple. Most insect models to date are linear in the system variables with highly complicated environmental forcing functions. This chapter will examine simple models that are nonlinear in the system variables with and without environmental forcing.

Describing insect population dynamics with models is not an easy task. Along with difficulties of measurement there are some serious problems inherent in the study of field populations. As Schaffer (1985) has pointed out, these systems are always embedded in a larger physical and biological environment, and the dynamics and equilibria will reflect this embedding (and will therefore depend on how the system boundaries are drawn). As careful as one might be in selecting boundaries, isolation is never quite complete and there is always "noise". Since it represents an unexplained erosion of predictability, noise has been traditionally interpreted as random error, and that apparent randomness can occur in simple deterministic systems has only recently been appreciated (Lorenz, 1963; May, 1974a; Olsen and Degn, 1985). Some ways in which this might occur in insect population models and perhaps even real systems will be discussed here.

The term "chaos" has been used to describe random-looking dynamics from a deterministic model. Chaos can be defined loosely as aperiodic dynamics from a deterministic system which is sensitive to initial conditions such that arbitrarily close initial points will diverge from one another at an exponential rate. Initial conditions can never be measured or even specified to infinite accuracy, and this means that no chaotic system can be predicted for long without updating the model from the real system (even if one has a "perfect" model). Very simple nonlinear models can have chaotic dynamics (Crutchfield et al., 1986; Holden, 1986; Jensen, 1987). Some of the simplest of these are discrete generation population models, reminiscent of univoltine insects (May, 1974a; 1976; May and Oster, 1976). This says in effect that even simple ecological systems could be chaotic, and in fact several human diseases appear to actually have chaotic dynamics (Schaffer, 1985; Schaffer and Kot, 1985; Olsen et al., 1988; Olsen and Schaffer, 1990).

While complicated dynamics can occur in very simple systems, certain key elements seem to be either necessary for or conducive to chaotic behavior. Typical among these are delayed nonlinear feedback (Hoppensteadt, 1979; Sparrow, 1980, 1981; Holden and Muhamad, 1986) and/or periodic forcing of nonlinear systems (Glass et al., 1983, 1984; Aron and Schwartz, 1984; Inoue and Kamifukumoto, 1984; Schaffer and Kot, 1986), in addition, high dimension seems to encourage chaotic dynamics. Thus, nonlinear systems with delayed feedback, environmental forcing and high dimension are potential candidates for chaotic behavior. Virtually all biological systems fit this description, and it has already been shown that many biochemical and physiological models and the systems they represent can be chaotic (Mackey and Glass, 1977; Williamowski and Rössler, 1980; Simoyi et al., 1982; Epstein, 1983; Holden and Muhamad, 1984; Pool, 1989). Field populations of insects are more elusive systems to study however and do not yield their secrets as easily. Consequently we will adopt the technique of adding small amounts of "realism" to simple models. The implications of the effects on the models will then be discussed.

Simple Models with Complex Dynamics: Examples

The simplest population model with nonlinear delayed feedback is the discrete generation logistic growth equation:

$$N_{t+1} = N_t + rN_t\left(1 + \frac{N_t}{k}\right) \qquad (3.1)$$

where

N = population density
r = reproductive rate per generation
K = environmental carrying capacity in density units

The substitution N = xK(1 + r)/r produces the famous one-parameter logistic map

$$x_{t+1} = ax_t(1 - x_t) \tag{3.2}$$

where a is (1 + r), (0 < a < 4), and x is dimensionless being defined by the substitution given above (0 < x < 1).

The dynamic complexity of this simple-looking model is suggested by a massive literature on the subject (May, 1974a, 1975, 1976, 1978a; May and Oster, 1976; Guckenheimer et al., 1977; Oster, 1978; Feigenbaum, 1983; Olsen and Degn, 1985; Lauwerier, 1986; Devaney, 1987; Jensen, 1987). It is scientifically embarrassing that the complex dynamics of equation 3.1 and its close relatives escaped notice until May's classical papers uncovered them. The implications are very fundamental for all of science: If deterministic models as simple as equation 3.2 can be chaotic having apparently random behavior, then what can we expect from the complexity of the real world?

Models like equation 3.1 where

$$x_{t+1} = f(x_t) \tag{3.3}$$

with $f(x_t)$ being a "one-hump" nonlinear function (or even piecewise linear) usually share a common type of approach to chaos as parameters change (Feigenbaum, 1983).

For equation 3.2 the equilibrium point (1 – 1/a) is attracting for (1 < a < 3). For (3 < a < 3.57) one observes a cascade of attracting cycles containing 2, 4, 8, 16, . . . 2^k points, (a "bifurcation sequence") the cycles increasing in length and complexity as a is increased. When (3.57 < a < 4) one finds chaos and cycles of all periods (including infinite periods) depending upon the value of a.

The sensitivity of equation 3.2 to initial conditions is illustrated in figure 3.1 where shifting the initial value of x in equation 3.2 by only 10^{-10} causes the trajectory to be completely different in about 50 iterations. Consequently there is no way to simulate a real chaotic system accurately for very long. Inaccuracy in initial conditions (no matter how small) will grow exponentially with time and the simulation will soon be following a different trajectory than the actual system. The weather seems to behave this way and is therefore a suspicious candidate for a chaotic system.

It is therefore no surprise that chaos was first discovered in a simplified weather model which is a good example of chaos in a system of differential equations. This system, given by:

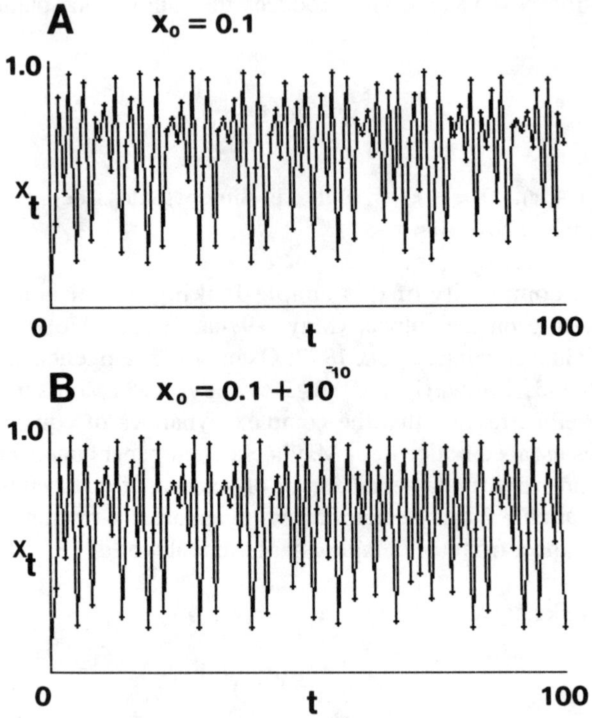

Figure 3.1–Divergent trajectories produced from equation 3.2 for a = 3.8 for initial conditions (x_0) which differ by only 10^{-10}.

$$\dot{x} = \sigma(y - x)$$
$$\dot{y} = rx - y - xz$$
$$\dot{z} = xy - bx \tag{3.4}$$

was introduced by Lorenz (1963, 1979) as a simplified relation between three weather variables where x = rate of convective transport, y = horizontal temperature gradient, z = vertical temperature gradient, and $\dot{x} = dx/dt$ is used to represent the time derivative (σ, r, and b are constants). The dynamic behavior of this system is so complex that it has been the subject of a book (Sparrow, 1982), a "Nova" television special (*Chaos*, Jan. 1989) and hundreds of research papers.

As a final example, consider a discrete time system with periodic forcing. This type of model has been used to represent an oscillatory system with periodic forcing such as the heart-pacemaker system (Keener and Glass, 1984), charged particles in electric fields (Chirikov, 1979) or with small modification, a cyclic population with seasonal forcing as suggested below. This is often written as:

$$x_{t+1} = ax_t + B\sin(2\pi y_t)$$
$$y_{t+1} = y_t + x_{t+1} \qquad (3.5)$$

where

x = the angular velocity of the system
y = the angular position of the system with
A = the damping coefficient, A = 1 implying a frictionless system
B = the amplitude of the periodic input

(Short and York, 1984; Jensen, 1987)

Forcing the system harder (increasing B) and reducing friction (A→1) will destabilize the system. This type of system has periodic, quasiperiodic (almost periodic) or chaotic behavior depending on initial conditions and the parameters A and B. The size of the chaotic region of initial conditions in (x, y) space increases with the amplitude of forcing, B (Jensen, 1987).

Distributed Delays with Nonlinear Feedback

Consider the delayed feedback system illustrated in figure 3.2. This system can represent a population model with a distributed developmental delay followed by a nonlinear reproductive feedback, $f(x_n)$. This can be written as:

$$\dot{x}_1 = [f(x_n) - x_1]\frac{n}{\tau}$$
$$\dot{x}_i = (x_{i-1} - x_i)\frac{n}{\tau}, \qquad i = 2, \ldots, n \qquad (3.6)$$

where

x = population density
n/τ = the rate of movement through the n-stage distributed delay
τ = the mean time in the delay

It has been shown by Sparrow (1980) that this standard feedback system with $f(x_n) = rx_n\exp(-x_n)$ and n = 50 goes through a bifurcation route to chaos just like discrete time systems represented by equation 3.3. This can be understood by considering the open-loop system where the feedback is simply an input to the linear distributed delay. In Laplace transform notation the transfer function of the delay is $G(s) = (1 + \tau s/n)^{-n}$ (Manetsch, 1976). In the Laplace transformed system (eq. 3.6) the output from the delay $[(n/\tau)x_n]$ is equal to the transfer function times the input $[(n/\tau)f(x_n)]$ or:

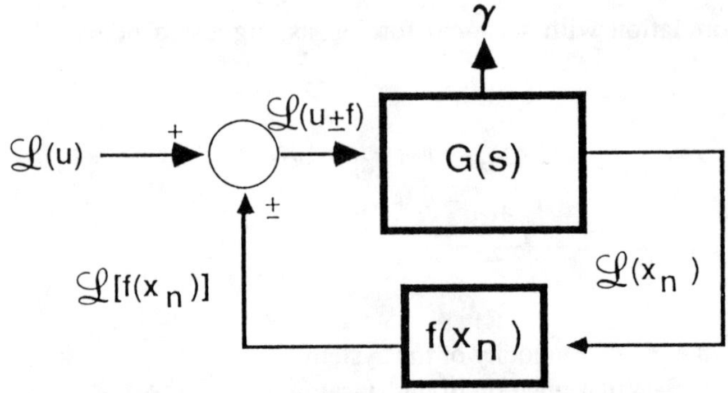

Figure 3.2–Population feedback loop having a distributed developmental delay with transfer function G(s), and nonlinear reproductive feedback function, f(x). u is an external forcing function, and γ is a distributed flow during the delay process. $\mathcal{L}$ (•) indicates a Laplace transformed variable.

$$\left(\frac{n}{\tau}\right)\mathcal{L}(x_n) = \frac{n}{\tau} G(s)\, \mathcal{L}[f(x_n)] \tag{3.7}$$

where $\mathcal{L}(\cdot)$ is the Laplace operator.

Cancelling the common factor (n/τ) and letting n become large, we have:

$$\lim_{n\to\infty}\left(1+\frac{\tau s}{n}\right)^{-n} = e^{-\tau s} = G(s) \tag{3.8}$$

and since $e^{-\tau s}$ is the transfer function of a discrete delay:

$$\mathcal{L}[x(t)] = e^{-\tau s}\, \mathcal{L}[f(x(t))]$$

$$\mathcal{L}[x(t)] = \mathcal{L}[f(x(t-\tau))] \tag{3.9}$$

so
$$x(t) = f(x(t-\tau))$$

and if we let τ arbitrarily be equal to one generation, we have difference equations, $x_{t+1} = f(x_t)$, like equation 3.3 which can be chaotic. Thus Sparrow's example shows that continuous systems like equation 3.6 can be chaotic, and that increasing dimension alone in this type of system can cause chaos.

Stability of the continuous system, equation 3.6, near equilibrium requires that the roots (λ) of its characteristic equation:

$$\lambda = \left[\left(\pm \overset{*}{f}\right)^{\frac{1}{n}} - 1\right]\frac{n}{\tau} \tag{3.10}$$

have negative real parts where $\overset{*}{f} = df/dx \mid \overset{*}{x}$, and $\overset{*}{x}_i = \overset{*}{x}$ is the equilibrium point for all i. [Compare Savageau (1976) for an example from a biochemical model]. $\overset{*}{f}$ can be ± to cover the possibility of positive or negative feedback slope at equilibrium since either can occur. Requiring that the largest λ-value have a negative real part, one arrives at the stability condition:

$$-\sec^n\left(\frac{\pi}{n}\right) < \overset{*}{f} \tag{3.11}$$

For a high dimensional delay (n large) this collapses to $-1 < \overset{*}{f} < 1$ which is the difference equation result for systems like equation 3.3. Thus the stable range for $\overset{*}{f}$ shrinks with increasing n, and we have an increasing likelihood of chaos as n becomes larger.

The Effects of Seasonal Periodic Forcing

The weather is an approximate kind of periodic forcing that needs to be considered in insect models. Discrete time models like equations 3.2 and 3.3 can be periodically forced by considering a spring and fall breeding (bivoltine) insect with nonoverlapping generations having different reproductive rates in the two seasons (Kot and Schaffer, 1984). These authors used an extension of equation 3.2 to represent each season:

$$x_f = \alpha x_s (1 - x_s)$$
$$x_s = \beta x_f (1 - x_f) \tag{3.12}$$

where

subscript s	=	the spring population
subscript f	=	the fall population
α	=	the reproductive rate (spring)
β	=	the reproductive rate (fall)

This seasonal form can be either periodic or chaotic depending upon the deviation between α and β when the mean geometric reproductive rate ($\sqrt{\alpha\beta}$) is only 2.5 while the nonseasonal equation 3.2 requires a > 3.0 for cycles and a > 3.57 for chaos. In addition, different attracting behaviors ranging from stable points to chaos can coexist in this seasonal form and the system can settle into one or the other or move back and forth between them depending upon initial conditions.

Periodic forcing can also produce chaotic dynamics in continuous differential equations with only two variables and no time lags. Inoue and Kamifukumoto (1984) have shown that the Lotka-Volterra predator-prey model can be chaotic if the prey's reproductive rate is subjected to sinusoidal forcing. If, in addition, a time lag is included in the predator's reproduction, then multiple attracting dynamic behaviors can coexist, the outcome depending only upon initial conditions (Allen, 1989). Epidemic models with constant coefficients which normally have attracting point stable behavior can be chaotic when disease transmission rate is made periodic (as it would tend to be in a seasonal environment) (Schwartz and Smith, 1983; Aron and Schwartz, 1984). Apparent chaos in childhood disease epidemics has been demonstrated from data sets by Schaffer and Kot (1985) and Olsen et al. (1988) where one-dimensional discrete time models of the form of equation 3.3 were derived from the data using a method of Takens (1981).

Now reconsider the feedback system (eq. 3.5) with an additive periodic forcing in the first stage of the delay and a flow, $\pm\gamma$ (this could represent food input or mortality output) evenly distributed along the delay chain. Equation 3.5 is modified to:

$$\dot{x}_1 = [f(x_n) - x_1]\frac{n}{\tau} \pm \gamma x_1 + u(t)$$

$$\dot{x}_i = (x_{i-1} - x_i)\frac{n}{\tau} \pm \gamma x_i, \qquad i = 2, \ldots, n \tag{3.13}$$

For the moment ignore the periodic forcing [let $u(t) = 0$] and consider the effect of the distributed flow $\pm\gamma$ alone. With some assistance from Vansickle (1977), the transfer function is modified to $G(s) = [1 + \tau(s \pm \gamma) / n]^{-n}$, and after some algebra, a modification of equation 3.11 emerges as the new stability condition:

$$-\sec^n\left(\frac{\pi}{n}\right) < \left[\left(1 \pm \frac{\gamma\tau}{n}\right)^{-n}\right]\overset{*}{f} < 1 \tag{3.14}$$

This condition indicates that γ is strongly destabilizing for values near $\gamma = -n/\tau$. Then $\overset{*}{f}$ is multiplied by a large value which tends to push it out of the stable range in equation 3.14. For values of γ far from $-n/\tau$, the distributed flow becomes stabilizing, multiplying $\overset{*}{f}$ by a small value. When $\gamma = 0$ we have no distributed flow and equation 3.14 becomes equation 3.11, and for non-zero γ and large n we have the difference equation (discrete delay) case:

$$-1 < (e^{\pm\gamma\tau})\overset{*}{f} < 1 \tag{3.15}$$

where loss ($-\gamma$) is stabilizing and gain ($+\gamma$) is destabilizing.

Thus, a simple constant flow that is evenly distributed during development effects dynamics and can even cause instability and chaos.

Returning now to the periodic forcing problem, we can follow the open loop transfer function method which led to equation 3.9 and obtain:

$$\mathcal{L}(x) = G(s)\ \mathcal{L}(u + f) \tag{3.16}$$

and since:

$$\lim_{n \to \infty} \left[1 + \frac{\tau(s \pm \gamma)}{n}\right]^{-n} = e^{\pm\tau\gamma} e^{-\tau s} \tag{3.17}$$

we have for large n:

$$\mathcal{L}[x(t)\ ] = e^{\pm\tau\gamma} \mathcal{L}[f(x(t - \tau) + u(t - \tau))] \tag{3.18}$$

and so as before we obtain a difference equation:

$$x(t - \tau) = e^{\pm\tau\gamma}[f(t - \tau) + u(t - \tau)] \tag{3.19}$$

If we ignore the distributed flow (let $\gamma = 0$), and look only at the periodic forcing, we find the discrete time system:

$$x_{t+\tau} = f(x_t) + u_t \tag{3.20}$$

and since u(t) is periodic, we let $u_t = A\cos 2\pi\phi_t$, and obtain the two-variable system:

$$x_{t+\tau} = f(x_t) + A\cos 2\pi\phi_t$$

$$\phi_{t+\tau} = \phi_t + \frac{\tau}{T} \tag{3.21}$$

where T is the period of u(t).

The dynamics of this system are considerably more complicated than its unforced counterpart, equation 3.3. Models from cardiac dynamics (Glass et al., 1983, 1984, 1986; Bélair and Glass, 1985) and the forced pendulum [Chirikov, 1979; Jensen, 1987; (eq. 3.5)] are similar. Analysis of equation 3.21 when $f(x_t)$ is equation 3.2 indicates that the system is attracted to an invariant curve on a cylinder (wrapping ϕ on itself gives rise to a circle while x can move "up and down" creating a circular surface, i.e., a cylinder) (Kevrekidis et al., 1985), and at the onset of chaos the cylinder becomes a fractal structure (Kaneko, 1984). Chaos is induced at lower values of the parameter of f than in equation 3.2 without forcing, and chaos can also be induced by varying the amplitude of the forcing (A).

The principal result here is that when a feedback system (a population) such as figure 3.2 and equation 3.6 is forced by the environment, stability is greatly affected. Constant distributed flows can be either stabilizing or destabilizing. Periodic forcing is almost always destabilizing and can produce chaos in models when none would occur otherwise. Thus, in insect populations, "outside" influences which are always present can induce chaos in otherwise stable or periodic systems.

Are Large Systems More Easily Chaotic?

We have seen that increasing the dimension of the delay process in a nonlinear feedback model tends to reduce stability and may actually cause chaotic dynamics. The idea that larger systems tend to be less stable has been expressed by Mees (1981) who stated that the "folklore" supports this notion. It is true that the types of complex motion which can arise increases with dimension (number of variables). In continuous systems, chaos can occur in a two-dimensional system only when periodic forcing is applied. When we have a three-dimensional system, chaos can occur without forcing or time lags (Gilpin, 1979). In four-dimensions we can have "hyperchaos" or the coexistence of two chaotic attractors in which the system switches back and forth between them (Rössler, 1979), and a whole hierarchy of increasingly complex behaviors is possible as the number of state variables (dimension) increases (Rössler, 1982).

In the pursuit of stability results for more complex systems, let us extend the single population loop of figure 3.2 to a "loop of loops" system as pictured in figure 3.3. In this system each species receives input from the previous one in the chain and passes its output on to the next, a bit like links in a trophic chain. Overall feedback, h, is a small feedback of the last species on the first perhaps via its posthumous remains. In the interest of simplicity, we assume that each species has identical parameters so that near equilibrium each loop has transfer function:

$$\hat{T} = \frac{G}{\left(1 \pm \overset{*}{f} G\right)} = \left(G^{-1} \pm \overset{*}{f}\right)^{-1} \tag{3.22}$$

where G is $(1 + \tau s/n)^{-1}$ as before.

Since the open loop transfer function for a whole chain of N such loops is $\hat{T}^N$, then the closed loop transfer function for the multiple loop system of N species is:

$$T = \frac{\hat{T}^N}{\left(1 \pm h\,\hat{T}^N\right)} \tag{3.23}$$

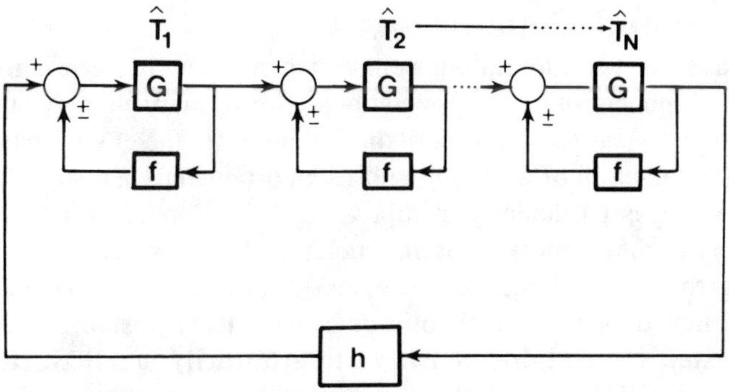

Figure 3.3–Loop of loops feedback system with component loops like those of figure 3.2. $\hat{T}_i$ is the transfer function of the *ith* component loop. The total system is *less* stable than the single loop subsystems.

From the denominator the characteristic equation is:

$$\left[\left(1 \pm \frac{\tau s}{n}\right)^n \pm \overset{*}{f}\right]^N \pm h = 0 \tag{3.24}$$

whose roots must have negative real parts for stable behavior.

Letting N (the number of species) be large and doing some algebra, one arrives at the stability criterion for this higher dimension system as:

$$1 - \sec^n\left(\frac{\pi}{n}\right) < \overset{*}{f} < 1 \tag{3.25}$$

Since the left hand side has been increased by 1 over the single species counterpart, equation 3.11, this system has a narrower stable range for $\overset{*}{f}$, and in that sense the higher dimension system is less stable than the single species version. If N were not very large, we would still intuitively expect a lower stable limit on $\overset{*}{f}$ between the single and multiple loop results in equations 3.11 and 3.25. This decline in stability is in general agreement with other results which indicate decreased stability as the number of variables and connections in a system increase (Gardener and Ashby, 1970; May, 1972, 1974b, 1978b; McMurtie, 1975; Tregonning and Roberts, 1979). Thus, theory indicates that large complex systems are inherently less stable than simpler ones, and the possibility of chaotic behavior seems to be greater in complex high dimension systems.

Summary and Conclusions

"Chaos" or pseudorandom behavior can occur in surprisingly simple deterministic models of insect populations. The primary ingredients for chaos are delayed nonlinear feedback, periodic forcing from the environment and high dimension. Interaction of a system with its environment via distributed flows or periodic forcing is particularly disruptive, and developmental delays are more destabilizing as they approach discrete delays. These results seem to imply that increasing the level of realism in insect population models increases the likelihood that they will be chaotic or at very least unstable. Additionally, chaotic systems are highly sensitive to arbitrarily small shifts in initial conditions (fig. 3.1), so that a small perturbation at any time results in a large change later on.

These results have very fundamental and rather disturbing implications. If simple, completely deterministic systems can have extremely complex random-looking dynamics, and if increasing amounts of realism increase the likelihood of this behavior, then we should be prepared for the possibility that insect populations may not be predictable for very long (even if we have a "perfect" model of the system). There is some scientific consolation in the fact that fairly simple models can produce this kind of dynamics. The fact remains, however, that if insect populations are chaotic, then models of these systems must be continually updated to prevent exponential divergence of the model and the real system.

References

Allen, J. C. 1989. Are natural enemy populations chaotic? In *Estimation and Analysis of Insect Populations*, ed. L. L. McDonald, J. A. Lockwood and J. A. Logan. New York: Springer.

———. 1988. Averaging functions in a variable environment: A second order approximation method. *Environmental Entomology* 17:621-625.

———. 1976. A modified sine-wave method for calculating degree-days. *Environmental Entomology* 5:388-396.

Aron, J. L. and I. B. Schwartz. 1984. Seasonality and period-doubling bifurcations in an epidemic model. *Journal of Theoretical Biology* 110:665-680.

Bélair, J. and L. Glass. 1985. Universality and self-similarity in the bifurcations of circle maps. *Physica* 16D:143-154.

Chirikov, B. V. 1979. A universal instability of many dimensional oscillator systems. *Physics Reports* 52:264-379.

Crutchfield, J. P., J. D. Farmer, N. H. Packard and R. S. Shaw. 1986. Chaos. *Scientific American* 255:46-57.

Devaney, R. 1987. *An Introduction to Chaotic Dynamical Systems*. Reading, MA: Addison-Wesley Publ. Co.

Epstein, I. R. 1983. Oscillations and chaos in chemical systems. *Physica* 7D:47-56.

Feigenbaum, M. J. 1983. Universal behavior in nonlinear systems. *Physica* 7D:16-39.

Gardener, M. R. and W. R. Ashby. 1970. Connectance of large dynamical (cybernetic) systems: Critical values for stability. *Nature* 288:784.

Gilpin, M. E. 1979. Spiral chaos in a predator-prey model. *American Naturalist* 113:306-308.

Glass, L., A. Shrier and J. Bélair. 1986. Chaotic cardiac rhythms. In *Chaos*, ed. A. V. Holden. Princeton, NJ: Princeton Univ. Press.

Glass, L., M. R. Guevera, J. Bélair and A. Shrier. 1984. Global bifurcations of a periodically forced biological oscillator. *Physics Review* 29A:1348-1357.

Glass, L., M. R. Guevera, A. Shrier and R. Perez. 1983. Bifurcation and chaos in a periodically simulated cardiac/oscillator. *Physica* 7D:89-101.

Guckenheimer, J., G. Oster and A. Ipaktchi. 1977. Dynamics of density-dependent population models. *Journal of Math Biology* 4:101-147.

Holden, A. V., ed. 1986. *Chaos*. Princeton, NJ: Princeton Univ. Press.

Holden, A. V. and M. A. Muhamad. 1986. A graphical zoo of strange and peculiar attractors. In *Chaos*, ed. A. V. Holden. Princeton, NJ: Princeton Univ. Press.

Holden, A. V. and M. A. Muhamad. 1984. The identification of deterministic chaos in the activity of single neurones. *Journal of Electrophysiology Tech.* 11:135-147.

Hoppensteadt, F. C., ed. 1979. Nonlinear oscillations in biology. *Lectures in Applied Mathematics*, vol. 17. Providence, RI: Am. Mathematics Society.

Inoue, M. and H. Kamifukumoto. 1984. Scenarios leading to chaos in a forced Lotka-Volterra model. *Prog. Theory in Physics* 71:930-937.

Jensen, R. V. 1987. Classical chaos. *Am. Scientist* 75 :168-181.

Kaneko, L. 1984. Fractilization of a torus. *Prog. Theor. Phys.* 71:1112-1115.

Keener, J. P. and L. Glass. 1984. Global bifurcations of a periodically forced nonlinear oscillator. *J. Math. Biol.* 21:175-190.

Kevrekidis, I. G., R. Aris, L. D. Schmidt and S. Pelikan. 1985. Numerical computation of invariant circles of maps. *Physica* 16D:243-251.

Kot, M. and W. M. Schaffer. 1984. The effects of seasonality on discrete models of population growth. *Theor. Pop. Biol.* 26:340-360.

Lauwerier, H. A. 1986. One-dimensional iterative maps. In *Chaos*, ed. A. V. Holden. Princeton, NJ: Princeton Univ. Press.

Logan, J. A., D. A. Wolkind, S. C. Hoyt and L. K. Tanigoshi. 1976. An analytic model for description of temperature-dependent rate phenomena in arthropods. *Environmental Entomology* 5:1133-1140.

Lorenz, E. N. 1979. On the prevalence of aperiodicity in simple systems. In *Global Analysis*, ed. M. Grmela and J. E. E. Mardsen. Lect. Notes in Math. 755:53-75.

———. 1963. Deterministic nonperiodic flows. *Journal of Atmospheric Science* 20: 130-141.

Mackey, M. C. and L. Glass. 1977. Oscillation and chaos in physiological control systems. *Science* 197:287-289.

Manetsch, T. 1976. Time-varying distributed delays and their use in aggregative models of large systems. *IEEE Trans.* SMC-6:547-553.

May, R. M. 1978a. Mathematical aspects of the dynamics of animal populations. In *Studies in Mathematical Biology II: Populations and Communities*. Studies in Mathematics, vol. 16, ed. S. A. Levin. Providence, RI: Math. Assoc. Am.

———. 1978b. Patterns in multi-species communities. In *Theoretical Ecology*, ed. R. M. May, 197-227. Oxford, England: Blackwell.

———. 1976. Simple mathematical models with very complicated dynamics. *Nature* 261:459-467.

———. 1975. Biological populations obeying difference equations: Stable points, stable cycles and chaos. *J. Theor. Biol.* 51:511-524.

———. 1974a. Biological populations with non-overlapping generations: stable points, stable cycles and chaos. *Science* 186:645-647.

———. 1974b. *Stability and Complexity in Model Ecosystems*, 94-102. Princeton, NJ: Princeton Univ. Press.

———. 1972. Will a large complex system be stable? *Nature* 238:413-414.

May, R. M. and G. F. Oster. 1976. Bifurcations and dynamic complexity in simple ecological models. *American Naturalist* 110:573-599.

McMurtie, R. E. 1975. Determinants of stability of large randomly connected systems. *Journal of Theor. Biology* 50:1-11.

Mees, A. I. 1981. *Dynamics of Feedback Systems*. New York: Wiley & Sons.

Olsen, L. F. and H. Degn. 1985. Chaos in biological systems. *Q. Rev. Biophysics* 18:165-225.

Olsen, L. F. and W. M. Schaffer. 1990. Chaos versus noisy periodicity: Alternative hypotheses for childhood epidemics. *Science* 248:499-504.

Olsen, L. F., G. L. Truty and W. M. Schaffer. 1988. Oscillations and chaos in epidemics: Six childhood diseases in Copenhagen, Denmark. *Theor. Pop. Biology* 33:344-370.

Oster, G. 1978. The dynamics of nonlinear models with age structure. In *Studies in Mathematical Biology II: Populations and Communities. Studies in Mathematics*, Vol. 16, ed. S. A. Levin, Providence, RI: Math. Assoc. Am.

Pool, R. 1989. Is it healthy to be chaotic? *Science* 243:604-607.

Rapp, P. E. 1986. Oscillations and chaos in cellular metabolism and physiological systems. In *Chaos*, ed. A. V. Holden. Princeton, NJ: Princeton Univ. Press.

Rössler, O. E. 1982. The chaotic hierarchy. *Z. Naturforsch.* 38a:788-801.

———. 1979. Chaotic oscillations: An example of hyperchaos. In *Nonlinear Oscillations in Biology*. Lect. in Appl. Math., Vol. 17, ed. F. C. Hoppensteadt. Providence, RI: Am. Math. Soc.

Savageau, M. A. 1976. *Biochemical Systems Analysis,* 188-190. Reading, MA: Addison-Wesley, Publ. Co.

Schaffer, W. M. 1985. Order and chaos in ecological systems. *Ecology* 66:93-106.

Schaffer, W. M. and M. Kot. 1986. Differential systems in ecology and epidemiology. In *Chaos*, ed. A. V. Holden. Princeton, NJ: Princeton Univ. Press.

———. 1985. Nearly one-dimensional dynamics in an epidemic. *Journal of Theor. Biology* 112:403-427.

Schwartz, I. B. and H. L. Smith. 1983. Infinite subharmonic bifurcation in an SEIR epidemic model. *Journal of Mathematical Biology* 18:233-253.

Sharpe, P. J. H. and D. W. DeMichele. 1977. Reaction kinetics of poikilotherm development. *Journal of Theor. Biology* 64:649-670.

Short, T. and J. A. York. 1984. Truncated development of chaotic attractors in a map when the Jacobian is not small. In *Chaos and Statistical Methods*, ed. Y. Kuramoto. New York: Springer.

Simoyi, R. H., A. Wolf and H. L. Swinney. 1982. One-dimensional dynamics in a multicomponent chemical reaction. *Phys. Rev. Lett.* 49:245-248.

———. 1982. The Lorenz equations: Bifurcations, chaos, and strange attractors, Vol. 41. *Appl. Math. Sci.* New York: Springer.

———. 1981. Chaos in a three-dimensional single loop feedback system with a piecewise linear feedback function. *J. Math. Anal. Appl.* 83:275-291.

———. 1980. Bifurcation and chaotic behavior in simple feedback systems. *Journal of Theor. Biology* 83:93-105.

Stinner, R. E., A. P. Gutierrez and G. D. Butler, Jr. 1974. An algorithm for temperature-dependent growth rate simulation. *Canadian Entomology* 106:519-524.

Takens, F. 1981. Detecting strange attractors in turbulence. In *Lecture Notes in Mathematics*, ed. D. A. Rand and L. S. Young. New York: Springer.

Tregonning, K. and A. Roberts. 1979. Complex systems which evolve toward homeostasis. *Nature* 281:563-564.

Vansickle, J. 1977. Attrition in distributed delay models. *IEEE Trans.* SMC-7:635-638.

Wagner, T. J., Hsin-I Wu, P. J. H. Sharpe, R. M. Schoolfield and R. N. Coulson. 1984. Modeling insect development rates: A literature review and application of a biophysical model. *Annals of the Entomological Society of America* 77:208-220.

Williamowski, K. D. and D. E. Rössler. 1980. Irregular oscillations in a realistic abstract quadratic mass action system. *Z. Naturforsch* 35:317-318.

4

Artificial Intelligence Approaches to Modeling Insect Systems

Nicholas D. Stone

Introduction

In recent years, the field of artificial intelligence (AI) has been recognized for its potential contributions to agriculture. Most of this attention has focused on the use of expert systems, a class of AI programs that capture and use human expertise to solve complex problems within some defined domain. The successful application of expert systems to diagnostic problems in medicine, chemistry, structural design, and business have encouraged agricultural researchers to develop similar systems that help solve management problems in natural resources. Expert systems, because of their emphasis on heuristic knowledge and nonalgorithmic problem solving methods, represent a new class of computer decision-aids and an alternative to the quantitative computer-based approaches that have traditionally been used to address problems in agricultural management. Even when detailed quantitative knowledge is not available, expert systems can still successfully describe system behavior and recommend management options.

While expert systems will undoubtedly play a major role in the future of tactical management of agricultural systems, the direct application of AI principles and tools to systems analysis and modeling may offer much more. AI is the study of ways to make computers act intelligently (Winston, 1984). One primary focus in AI research has been to develop new ways to store and use information in a computer that mimic the same processes in the human mind. The resulting AI techniques for knowledge representation and search, however, can be taken out of context, and used for purposes other than mimicking human reasoning. A relatively small step, for example, is to apply these techniques to mimic animal reasoning and behavior. When examined objectively, AI programs, environments, and methodologies are simply another set of tools with which to model nature. To date, they constitute some of the most advanced programming languages and environments available for modeling and simulation.

This chapter describes some of the ways in which AI has and might expand the horizons of modelers. After defining some key terms, four categories of applications are discussed; 1) AI programs alone; 2) AI-enhanced simulation languages; 3) AI programs in combination with traditional simulation models and operations research techniques; and 4) AI alternatives to traditional simulation. In several instances, entomological examples are used to illustrate the concepts.

Definitions

Traditional systems analysis relies on mathematical descriptions of nature. Differential equation models, difference equation models, and statistical models all use mathematics as the representation scheme. Computer implementation of such models are required for speed, numerical solutions of complex models, and application of models to practical problems. However, the computer's role is clearly secondary to the mathematical representation of the model. As a result, computer programming has typically been considered an ancillary process in systems analysis, subordinate to the actual modeling process. The choice of a programming or simulation language was merely a function of how adaptable that language was to the form of the mathematical model.

In AI, on the other hand, computer programming is paramount. AI research is based on computer representations of nature using languages that are nonalgorithmic and symbolic. Furthermore, designing an AI system is inseparable from programming the system. Virtually the entire process occurs at a computer workstation, using a programming environment. The programming problem is the modeling problem, and consequently the knowledge engineer is much more concerned with computer programming constructs than with mathematics.

If systems scientists are to embrace AI and utilize its potential, they must accept computer-based tools for representing nature as the equals of mathematical descriptions of nature. To some extent, this has been made easier by the evolution of personal computers, computer-assisted software (CAD, CAM, CASE, etc.) and fourth-generation programming languages, tools that free the programmer to concentrate on design concepts rather than details of coding syntax. These packages make it easy and compelling to work at the computer. Still, many systems scientists would rather build the models and leave the programming problems to someone else. Those who do wish to take advantage of the computer-based modeling tools of AI must become computer programmers and learn the fundamentals of AI programming.

The AI programming task can generally be broken down into two questions: 1) how should the relevant knowledge be represented in the program; and 2) how should that knowledge be used to solve problems. These two questions correspond to AI notions of knowledge representation, search, and inference.

Knowledge Representation. Knowledge representation is the way that information is stored in the computer. Mathematical equations are one type of knowledge representation, maps and graphs another. Because AI deals primarily with nonalgorithmic knowledge, the knowledge representation schemes of AI are geared toward describing qualitative relationships, including both correlative and causative information. One of the clear lessons of AI research is that choosing a good representation can often make a problem's solution trivial (Winston, 1984). It is because knowledge representation is such a central issue in AI that many AI programs are referred to as knowledge-based systems. Some of the more common structures for representing knowledge are defined below.

Rules are by far the most common representation scheme in expert systems. They are conditional statements with an if...then structure. Each rule represents an independent logical relationship between the antecedents (on the if-side) and the consequents (on the then-side). Rules in knowledge-based systems are therefore distinct from conditional statements in conventional programming languages in which if...then statements occur in a known context within a larger piece of code. Rules define knowledge, how to make conclusions based on information and prior conclusions. In classical programming languages, if...then structures are logical switches controlling program flow. In practice, the two roles of if...then structures can merge together. Some rules in expert systems may direct the flow of reasoning (Woolley and Stone, 1987), while some if...then statements in traditional programs can be used to describe heuristics (Logan, 1988). The key difference is the independent nature of rules in a knowledge base versus the embedded nature of if...then statements in traditional programs.

An object is a dynamic structure in an AI program. Conceptually, objects are the units manipulated by an AI program; they represent the actors or components of the system one is trying to model. But objects are also the lowest level of a hierarchy of object classes. A class might be a type of insect, a beetle, and an instance of that class would be a specific beetle object. Information about objects in general are stored in the class description. For instance, knowledge about the family Coleoptera common to all beetles would be stored in the class description. Since all beetles undergo complete metamorphosis, the class, "beetles", in a knowledge-base could record that information by making the value of a class-variable named "metamorphosis" equal to "complete". Specific beetle species, for example, *Rodolia cardinalis*, defined later as objects and members of the class, "beetles", would inherit the information contained in its class description. The type of development of *R. cardinalis* would be accessed through its inherited variable, metamorphosis.

In object-oriented programming languages (Stefik and Bobrow, 1986), objects contain both data and associated attributes and procedures. Furthermore, objects may be created and destroyed dynamically during program execution. For example, individual insects may be conceived as actors in an ecosystem and modeled as objects. Each insect object would have attributes—like its age, sex, location, nutritional status, activity, etc.—and procedures or methods that allow it to respond to its surroundings and stimuli. Among such methods might be fly, eat, age-a-day, mate, and so on. Program execution occurs by objects sending messages to other objects. There are no direct procedure calls since all procedures are locally defined in objects themselves. Objects and object classes can be created using various techniques, but one of the most common implementations is a frame-based language.

Frames are structures that allow sets of related knowledge to be grouped together. Frames can contain knowledge relevant to an object, a situation, or some concept (Tanimoto, 1987). They contain knowledge in slots, just as data records in Pascal contain knowledge in fields. Each slot has some value or filler, which may be a numerical value, a rule, or even a procedure. In the example above, an insect frame would contain slots for each attribute and for each method associated with the insect. Frames can also be used to group related sets of rules together in the knowledge base of an expert system.

Search and Inference. Search is the process of choosing a path from among a set of constrained choices in order to reach a goal. Inference is a related process of creating new knowledge from existing knowledge, and its most common form in AI programs is deductive inference. Inference and search are related in that choosing a search path can employ inferencing methods, and inferencing schemes often rely on a search strategy to process the knowledge base. In rule-based expert systems, for example, the process of solving a problem involves making a succession of inferences in a way that quickly leads to a solution.

Search is a general concept. Systems analysts are already familiar with optimal search strategies like the simplex algorithm for solving linear programming problems or the recursive algorithms of dynamic programming. Because AI programs are not generally mathematical, optimality is difficult to define and other strategies are employed. Heuristic search, for example, improves on blind search strategies by employing rules or evaluation functions to choose which path seems most likely to lead to the goal. Because systems scientists are familiar with the use of search algorithms, they also can perceive the potential application of AI search strategies in solving problems that have eluded classical optimization.

AI Systems

Once systems scientists are familiar with the new representation schemes and programming tools that AI has to offer, they will naturally find uses for them in a systems approach to problem solving. Many novel applications of AI techniques to agricultural and experimental systems have already been discovered, and we are just at the beginning of the AI revolution in the natural sciences.

Stand-alone Knowledge-based Systems. Because expert systems are the predominant AI program type in use and being developed in agriculture today, it is appropriate to begin with a discussion of how expert systems can be used in systems analysis. Two contributions of expert systems to systems analysis are: 1) an expansion of systems science methodology to a new type of system, the knowledge system; 2) providing a new way to model systems in which understanding has not reached the level of mathematical abstraction.

Modeling a Knowledge System. Large-scale or broadly defined problems are generally not well suited to expert systems (Davis and King, 1984). Building management expert systems for agricultural systems is a potentially impossible task, therefore, because the problem is so broad. Other approaches have more severe limitations, however. Classically, one would approach the problem by focusing on the biological and physical elements of the system, modeling these as well as possible using mathematical models, and using mathematical programming techniques to optimize selected objective functions. This generally involves a linked economic model, since optimal in a management setting usually means maximizing profit. However, mathematical models of this type have had limited utility. These models are limited because simulation technology is difficult to deliver, and only a small subset of the biological processes in most agroecosystems are understood well enough to simulate with a mathematical model. The same might be said for the economic system.

Alternatively, one can perceive farm-level crop production itself as a system, in which the component parts are decisions, and the linkages between components are information (fig. 4.1). The information flowing into and from decisions exists in several types: choices and decisions previously made that affect the current decision; objective information, the facts and information available from databases; simulation model output, including prediction and economic analysis derived from computer models; and heuristic knowledge or expert opinion about how the decision ought to be made. The last type includes, for example, the expertise needed to interpret the complex results from a simulation model as well as more fundamental rules of thumb and judgement that fill in the gaps in our objective knowledge base.

Each node or decision in such a system is eventually modeled by a rule set devoted exclusively to the specific problem defined in the box. In this way, the primary obstacle to building a large scale expert system—that the problem be narrowly defined—is removed. Each rule set focuses on a discrete, well defined problem, yet the system as a whole can solve a variety of more complex problems by using its component systems appropriately. A decision network or knowledge system is created in which each node is modeled by a rule base module, and the catalog of decision nodes represents the range of problems the system can respond to. Once the user has identified a starting point within the decision network, the system chains through modular rule-bases to reach a conclusion.

An example of such a decision network, developed for an expert system to recommend pest control actions for insect pests in western forests, is shown in figure 4.2. The expert was J. L. Robertson, an entomologist with the U.S. Forest Service. Possible choices are denoted within the boxes. Lines indicate information required or possibly required in order to make each decision. The next stage in system development is to replace each rectangle with a rule-base. By using this systems approach to the pest management problem, relatively unstructured problems like, "my trees are turning yellow,

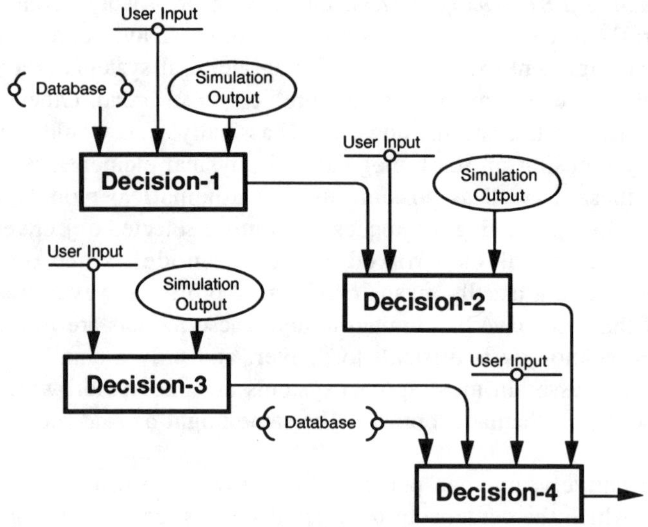

Figure 4.1–A knowledge system represented as a network of decisions linked by information flow. Four types of information are shown: user input of objective data, objective information stored in a database, simulation output, and previous decisions or choices made reflecting the opinion of experts.

what should I spray?" can be addressed by a series of well defined expert system modules.

Rule-based Modeling. When quantitative modeling is impossible or inappropriate, qualitative modeling techniques have sometimes been successful in capturing the essence of system behavior (Lendaris, 1980; Bonnicksen, 1982). Rule-based expert systems, although not developed with simulation in mind, offer another tool for qualitative simulation.

Holling (1978) classified modeling problems into four categories based on the extent of data and general understanding about the problem (fig. 4.3). Category A, in which there is considerable data but little understanding, is appropriate for statistical approaches. Category B, high in data and understanding, is the area best suited to quantitative models. Many mechanical and physical problems fit in this category and have been successfully modeled.

The problems faced in modeling insect populations, though, generally fall into categories C and D—little data and only incomplete understanding. When understanding is high enough in these cases (category D), it should be possible to capture that understanding in the form of a rule base describing how the system functions. In effect, one would be building an expert system that mimics the laws of the natural system. Instead of reasoning to some problem solution, such an expert system would reason about how the state of the system changes with time. This process would be event-driven, rather than time-driven, since changes would occur only as a result of deductions or events perceived by the system. Figure 4.4 shows a rule-based simulator, emphasizing that the state of

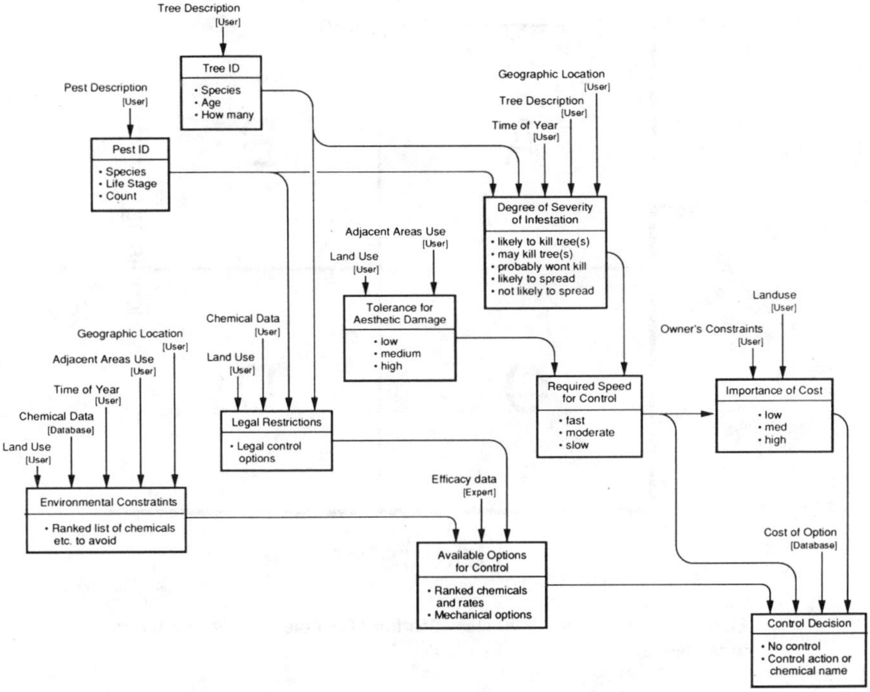

Figure 4.2–A function model for an expert system to recommend control decision for insect infestations in western forests. Key choices and decisions are shown in boxes; arrows represent information flow.

the world as understood by the inference engine is also the simulated state of nature. That state is updated through an inference procedure based on the evaluation of rules about biological system processes.

As an example, suppose one wanted to model the pattern of insecticide resistance in *Heliothis virescens* during the course of a cotton production season. While quantitative data are lacking, there is now a good understanding of how pyrethroid insecticide use in the early season causes a buildup of the resistant gene frequency and how the gene frequency diminishes once pyrethroid use is halted. Experts familiar with this pest also have some qualitative understanding of *Heliothis* movement behavior in relation to cotton phenology and growth. One can create the knowledge base for an expert system that will make conclusions and ask questions about events (like whether pyrethroid use has discontinued), by describing this knowledge in the form of rules, for example:

If cotton fields within a local region are fruiting,
then long range migration is minimal.

and

If long range migration is minimal and
pyrethroid use is discontinued.

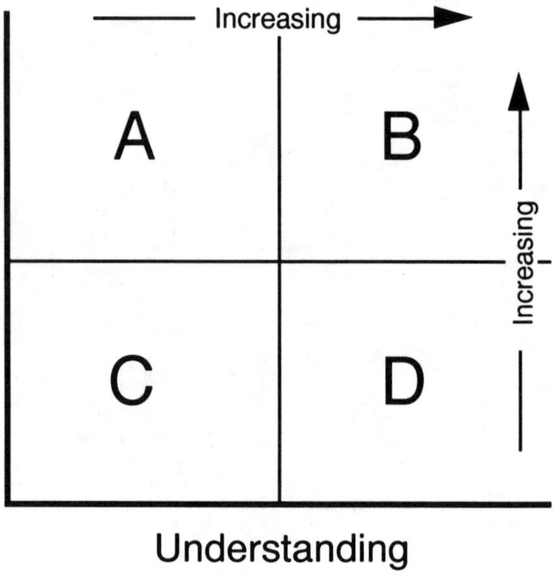

Figure 4.3–Holling's (1978) classification of problem types in relation to modeling.

then resistance gene frequency will decrease.

Such a model can make an immediate impact on management practices even though the data required to build a quantitative model may take years to collect. Furthermore, building this type of model formalizes the knowledge by which management decisions are made, and thereby makes such decisions more consistent and open to scrutiny.

Starfield and Bleloch (1986) have built a simple example of such an event-driven simulator used for minimizing brush fires in high hazard areas. Starfield has also begun work on a rule based simulation of an estuary system to allow managers to make informed decisions about salinity management (A. M. Starfield, personal communication).

AI-enhanced Simulation Languages. A number of researchers have pointed out that AI has the potential to greatly improve modelers' capabilities to model efficiently and effectively (Loehle, 1987; Logan, 1988). AI programs like expert systems are being constructed to aid in interpretation of raw biological data, to facilitate model design (G. Curry, personal communication), to choose appropriate mathematical functions for biological processes (Logan, 1988), and related programs may soon perform sensitivity analysis, iterative parameterization, and balancing of units in computational algorithms (Loehle, 1987).

Currently, the available AI tools to expedite traditional modeling are minimal. The most remarkable advances have been AI programs that symbolically manipulate mathematical formulae. Programs like Maxima™

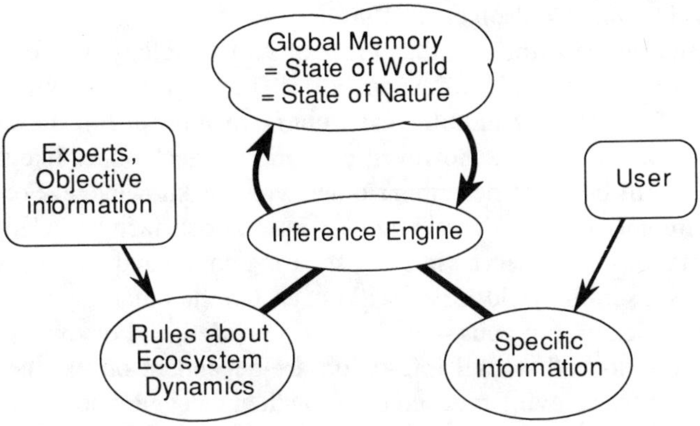

Figure 4.4–Diagram of a rule-based simulator. Knowledge base consists of rules describing system dynamics. User provides specific information. As the system makes deductions and learns new facts, the state of nature as reflected in the system's global memory or state of the world is updated.

(Pavelle et al., 1981), for instance, have been able to solve differential equations, compute definite and indefinite integrals, and simplify expressions at a level superior to human mathematicians. Even at the personal computer level, symbolic math programs exist today that can greatly simplify mathematical model construction and evaluation (e.g., MathCAD, DERIVE, Metrimatica, etc.).

The first steps toward AI-guided modeling have been taken, however, and it may not be long before one can build an insect model by using an interactive computer program to select appropriate process models, suggest experiments to provide useful data, generate computer code, and perform sensitivity and stability analyses. Still, the process of modeling will not be a good problem for AI until some consensus is reached on what makes a good model for any particular problem. As long as each model must be tailored to the specific application, every model will have to be built differently, and the process involved will remain very much an art. A more reasonable goal for AI would be to provide smart tools to modelers to make their work easier once an approach is decided upon.

Integrated AI/Traditional Applications. One of the most intense interests today in AI in agriculture has been integrating AI with existing applications. Expert system interfaces to complex systems, for example, have been developed to aid in monitoring experimental growth chambers (Jones and Haldeman, 1986); as front and back-ends to simulation models to allow them to be applied to practical problems like water and fertility management and pest control in cotton (Lemmon, 1986; Stone et al., 1987); and as heuristic enhancements to simulation models (Stone et al., 1986; Jones et al., 1985; Starfield and Bleloch, 1986). In addition, AI shells have been used to integrate various types of information systems into a new generation of decision support systems or expert support systems for management of natural resources (Naegele et al., 1986; Coulson et al., 1987).

It should not be surprising that a mix of AI technology with conventional modeling techniques could improve both. The symbolic, nonalgorithmic techniques of AI clearly enable researchers to incorporate new kinds of knowledge into their models. However, conventional techniques like simulation modeling are still better at describing other types of knowledge. For example, rules are extremely good at describing correlation or surface knowledge: if boll weevil infestations are observed in early-squaring cotton and boll weevil are not controlled, then serious yield losses will occur. On the other hand, they are not well suited to describing causative or deep knowledge. For example, cotton yield losses due to boll weevil infestation are dependent on weather, rates of development of the weevil larvae, cotton growth and development, temperatures in the field at the soil surface, humidity, and several other factors. Writing rules to describe such relationships would be tedious and inexact. If available, mathematical models are better suited to describing such deep knowledge.

Three types of application show considerable promise in improving the scope of traditional systems analysis: 1) heuristically enhanced models; 2) integrated expert systems; and 3) landscape simulators.

Heuristic Enhancement. Starfield and Bleloch (1986) make the sobering remark, "we never have all the data we want". Similar thoughts are never far from any modeler's mind. The classical approach to problems of missing or ambiguous data is to assume some default or to fit models to specific situations through iterative parameterization. AI techniques provide an alternative: where mathematics fails, use nonalgorithmic options. Thus, in their rule-enhanced model of a Kudu population in African game reserves, Starfield and Bleloch (1986) used a rule-based system to estimate departure from expected calf mortality rates based on expert experience. They showed that while a purely quantitative model could predict Kudu populations well overall, the enhanced model performed much better in years of severe weather.

Jones et al. (1985) similarly used expert-derived rules in an otherwise quantitative analysis of the control of velvetbean caterpillar in soybean. In part of their analysis, a rule-based system acted to improve the searching efficiency of an algorithmic optimization analysis to determine the cheapest insecticide available that would achieve some required kill. In effect, a heuristic search procedure was implemented through a rule-based expert system so that only certain likely candidate insecticides were considered in the simulation analysis.

Integrated Expert Systems. The term integrated expert systems has been used (Coulson et al., 1987) to describe knowledge-based systems that use integrated component expert systems to model management systems superimposed on deep models of natural resource systems. In effect two models of the managed system are needed, a management system model, and a natural system model. Systems containing these two elements to some degree include the COMAX system (Lemmon, 1986), the POMME system for pest management in apple orchards (Roach et al., 1986), the CALEX system for cotton management in California (Plant and Wilson, 1986), and the COTFLEX system for farm management in Texas (Stone et al., 1987).

The problem in designing integrated systems is to enable the appropriate knowledge source to provide information when it is needed and in the correct format. Simulation models, for example, require an input file and a routine to sort through the output to filter out the relevant information for decision making. In the rule-based component of integrated expert systems, triggering the simulation can be accomplished in several ways. For example, in the following rule, the value of cfh_population_trend (the cotton fleahopper population trend) is needed.

If cfh_sample is borderline and
not (cotton is close_to_next_phenological_stage) and
cfh_population_trend is increasing,
then recommended_action_for_cfh is control.

Assuming there is a computer model available that simulates cotton fleahopper population dynamics, another rule is needed to prepare the input file and trigger the model.

If cfh_input_ready = (call write_cfh_input_file[parameters...]) and
cfh_trend = (call cfh_simulation) and
cfh_trend is increasing,
then cfh_population_trend is increasing.

The syntax for calling external routines will vary among shells and operating systems, but the tasks accomplished by the external routines will remain the same: create an input file for the model based on knowledge in the expert system; call the model; and filter the output from the model so that the model returns the appropriate value to the expert system.

Landscape Simulation. Representation of heterogeneous landscapes has always been a problem for simulation modelers. Most insect-crop models assume a uniformly growing field (Stone and Gutierrez, 1986), or base area-wide predictions on generalizations from smaller, uniform systems (Hartstack et al., 1976). Still, spatial heterogeneity is extremely important in population ecology and has dramatic influence on the stability of small systems (Hassell and May, 1973).

Good tools for landscape representation do exist, however. One approach represents homogeneous regions within landscapes as objects in an object-oriented simulation; this approach is described in the next section. Another method uses a traditional computer representation scheme, a geographic information system (GIS), and is discussed below.

A GIS is a database management system for map-oriented data. For any region, a set of maps can be added to the database as digitized images to represent different data themes describing the region. Within each data theme, regions of homology are identified and attributes describing those regions are stored in tables that reference the map of the data theme. GISs then offer several features for manipulation of the graphical and tabular data. For instance, if data themes for field boundaries, crop, and soil type, were included in a GIS for a given farm, the GIS could quickly calculate and display on a map all regions within selected cotton fields planted in heavy clay soils. Alternatively, if weed histories were available in map form, all cotton fields planted in high risk nutsedge areas could be displayed to help farmers in early weed control decisions.

However, like any database, a GIS is static and contains no domain-specific knowledge to help its users manipulate its database. Recently, a few researchers (Robinson et al., 1987; White and Morse, 1987) have attempted to integrate GISs with expert systems to make the systems much more powerful. Such intelligent GISs (IGIS) could be used to answer management questions, as in the weed control example above, by including a knowledge base about how to spot potential trouble spots. Alternatively, if knowledge about ecosystem dynamics were added to a GIS, such a system could use its knowledge to make its data themes dynamic, thereby creating a landscape-level simulator.

Figure 4.5 shows an example from forest entomology of an IGIS under development in the Knowledge Engineering Lab (KEL) at Texas A&M. This study concerns southern pine beetle (SPB) infestations, which are known to be related to a forest stand hazard rating, lightning strikes, weather, and background insect populations (Coulson et al., 1986). The IGIS under development contains data about the eastern Texas forest landscape to permit calculation of hazard ratings for specific forest stands. Likelihood of southern pine beetle outbreaks is then predicted by the system. Spot growth of initial infestations hypothesized by the system can then be simulated by a model of SPB infestation growth (Feldman et al., 1981). Because this system is graphically oriented, it can provide forest managers with an illustrated simulation of SPB damage and spread in managed forests. As a management tool, such information will be invaluable.

AI-based Simulation. One form of AI simulation, rule-based, has already been discussed. Its main strength is in simulating systems in which quantitative data is lacking. A second approach is to use object-oriented programming (OOP) concepts (Stefik and Bobrow, 1986) and AI representation schemes both to model the environment as a set of objects, and to model populations as instances of a population or species class (Coulson et al., 1987).

Several advantages to this type of simulation are evident beyond the improved efficiency of data and procedure abstraction afforded by object-oriented languages. 1) Hybrid models (AI + conventional) can be built, because both algorithmic and symbolic knowledge can be associated with objects. A LISP method can take any form from a mathematical algorithm to a set of rules and an inference engine. 2) There is a direct conceptual correspondence between actors and interactions in the natural system and objects and methods, respectively, in the simulated system. For example, an insect object can send a plant object a message called biting to reflect the natural act of feeding. This makes programming and debugging very straightforward. 3) Heterogeneous landscapes can be modeled as sets of polygon objects linked by attributes describing their relative positions. 4) Simulation of populations at the individual level is possible. In OOP, individuals (objects) are the level of interest. Population equations are not appropriate. Instead, one attempts to model the

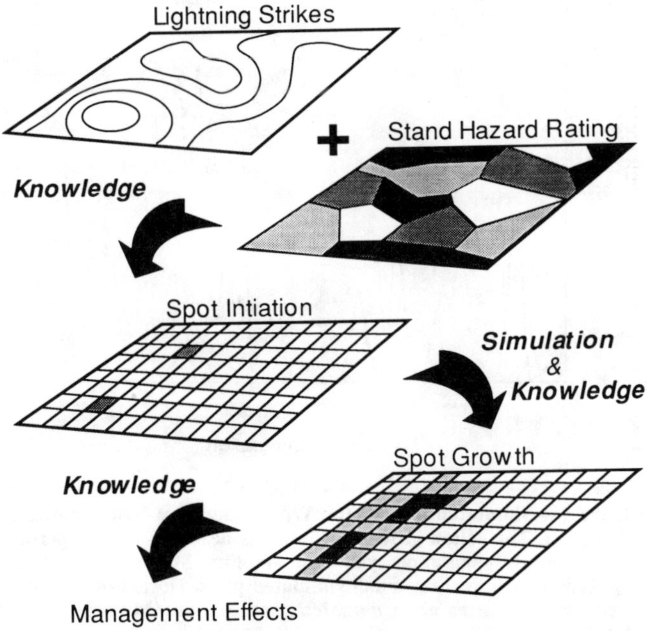

Figure 4.5–Landscape simulation using an intelligent geographic information system of southern pine beetle infestation and spread in relation to lightning strikes and forest stand hazard rating. Data themes are represented by large rectangles. Simulation output and heuristic knowledge are used by the system to update its data themes.

characteristics of individuals, the true actors in the environment, and to observe the net effects at the population level.

Of these, perhaps the most important to systems analysis and insect modeling is the last for two related reasons. First, by representing individuals as discrete instances of an object class, one can build behavioral models with relative ease. AI techniques are well suited to describing behaviors and the steps leading to behaviors. In fact, mimicking these processes in humans is one basic goal of AI. Methods that determine individual behavior and other methods that carry out specific behaviors can be associated with an object class representing some generic population unit (a species, or the males of a species, the adult males, etc.). Then, individuals can be created as instances of the object class, and these objects can interact. Behaviors need be defined only once in the object class. Each individual (object) will apply its inherited behaviors depending on its current and/or past experiences.

Second is the fact that population-level effects can be observed as emergent properties of this type of model. Using OOP models, one can test some of the basic assumptions about populations through simulation based on directly observable individual behaviors. For example, density dependent population growth is virtually always included in insect models without regard for the mechanism involved. Using an OOP approach, one could endow

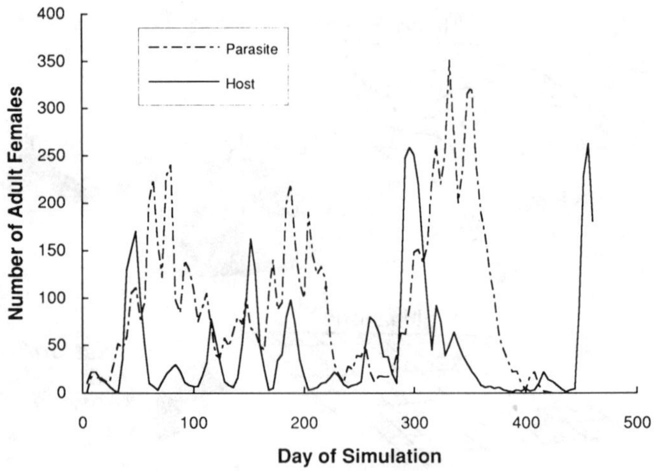

Figure 4.6–Simulation run of the WASP model of host-parasitoid interaction. The cotton growing season has been extended to permit examination of population cycles and stability. Shown are the total populations within three small simulated plots. Hosts are *Heliothis virescens*, parasites are *Campoletis sonorensis*. (Modified from Makela et al., 1988.)

individuals with the observed behaviors and traits that are assumed to explain density dependent population growth and then observe the behavior of the simulated population.

An OOP model of host/parasite population dynamics in cotton has recently been developed by M. E. Makela of the KEL at Texas A&M (Makela et al., 1988). The model simulates the interaction of *Heliothis virescens* and *Campoletis sonorensis* in cotton fields based on behavioral and phenological data. The simulated environment consists of three closely spaced cotton fields, each divided into regular parcel objects. Adult females of the two species are the other objects in the system, modeled as instances first of a generic female insect class with subclasses for the two species. Attributes of all female insects, e.g., ability to lay an egg and seek oviposition sites, are included in the generic class while more specific attributes and methods are modeled in the species-specific classes.

The model has been used to examine population dynamics and stability, dispersion patterns in the field, and density dependent effects. Figure 4.6 shows one simulation run over an artificially extended production season. Populations of host adult females, parasite females, and emigrating parasite females are shown over time. Population dynamics showed wide oscillations in which high *Heliothis* populations were quickly followed by a buildup of parasites and a crash of both. Examination of the model indicated that the stability of the model was due to local survival of hosts in relatively unparasitized parcels within the fields even when total parasite numbers were very high. Other interesting results of the simulations has been a pronounced edge-effect (despite the lack of any behavioral inclination toward the edges of fields), and an apparent mutual interference (Hassell and Varley, 1969) among the parasites at high densities despite the lack of any behavioral interference in the model. While the specific results may not be significant in relation to cotton production, the technique shows clear promise in helping to link observations about insect behavior more directly with population-level effects in the field.

Conclusions

New types of models, new approaches to simulation, and the ability to incorporate heuristic reasoning and qualitative elements into quantitative analyses are the main contributions thus far from AI to the field of systems analysis. In the future, intelligent modeling tools may make model building easier as well. Still another fundamental change that AI is bringing about is in the role of the computer. The computer is the work environment of AI. Programming cannot be the last step in model design and construction. Programming occurs throughout. In model conception, design, prototyping, knowledge engineering, and simulation, AI workstations and the tools they contain for knowledge representation, interactive simulation, debugging, and editing are essential.

Expert systems, currently the dominant AI application in agriculture, are here to stay. In capturing and delivering scarce expertise, they are an unmatched tool and the future successor to the Fact Sheet of the Extension Service. Still, expert systems' long-term influence on systems research is likely to be dwarfed by the more profound offerings of the field of AI.

Acknowledgments. The author wishes to thank members of the Knowledge Engineering Laboratory at Texas A&M, from whom much of the work presented here originates.

References

Bonnicksen, T. M. 1982. Biosocial systems analysis: An approach for assessing the consequences of resource policies. *Journal of Environmental Management* 15:47-61.

Coulson, R. N., L. J. Folse and D. K. Loh. 1987. Artificial intelligence and natural resource management. *Science* 237:262-267.

Coulson, R. N., R. O. Flamm, P. E. Pulley, T. L. Payne, E. J. Rykiel and T. L. Wagner. 1986. Response of the southern pine bark beetle guild (*Coleoptera: Scolytidae*) to host disturbance. *Environ. Entomol.* 15:850-8.

Davis, R. and J. King. 1984. The origin of rule-based systems in AI. In *Rule Based Expert Systems: the MYCIN Experiments of the Stanford Heuristic Programming Project,* eds. B. G. Buchannan and E. H. Shortliffe. Menlo Park, CA: Addison-Wesley Publishing Co.

Feldman, R. M., G. L. Curry and R. N. Coulson. 1981. A mathematical model of field populations of the southern pine beetle, *Dendroctonus frontalis*. *Ecological Modelling* 13:261-81.

Hartstack, A. W., J. A. Witz, J. P. Hollingsworth, R. L. Ridgeway and J. D. Lopez. 1976. MOTHZV-2: A computer simulation of *Heliothis zea* and *Heliothis virescens* population dynamics. USDA-ARS-S-127.

Hassell, M. P. and R. M. May. 1973. Stability in insect host-parasite models. *Journal of Animal Ecology* 42:693-726.

Hassell, M. P. and G. C. Varley. 1969. New inductive population model for insect parasites and its bearing on biological control. *Nature* 223:1133-1137.

Holling, C. S., ed. 1978. *Adaptive Environmental Assessment and Management.* Chichester: John Wiley & Sons.

Jones, J. W., P. Jones and P. A. Everett. 1985. Applying agricultural models using expert system concepts. ASAE Paper No. 85-5517. St. Joseph, MI: ASAE.

Jones, P. and J. Haldeman. 1986. Management of a crop research facility with a microcomputer-based expert system. *Transactions of the ASAE* 29(1):235-242.

Lemmon, H. E. 1986. Comax: An expert system for cotton crop management. *Science* 233:29-33.

Lendaris, G. G. 1980. Structural modeling–A tutorial guide. *IEEE Transactions of Systems, Man, Cybernetics* 10:807-40.

Loehle, C. 1987. Applying artificial intelligence techniques to ecological modelling. *Ecological Modelling* 37:192-212.

Logan, J. A. 1988. Toward an expert system for development of pest simulation models. *Environmental Entomology* 17:359-376.

Makela, M. E., S. B. Vinson and N. D. Stone. 1988. Host-parasitoid population dynamics in a heterogeneous environment. In *Proc. Soc. Computer Simulation Multiconference on Artificial Intelligence and Simulation: The Diversity of Applications,* 228-233. San Diego, CA: Society for Computer Simulation.

Naegele, J. A., R. N. Coulson, N. D. Stone and R. E. Frisbie. 1986. The use of expert systems to integrate and deliver IPM technology. In *Integrated Pest Management on Major Agricultural Systems,* eds. R. E. Frisbie and P. L. Adkisson, 692-711. Texas Agric. Exp. Sta. MP-1616.

Pavelle, R., M. Rothstein and J. Fitch. 1981. Computer algebra. *Scientific American* 245(6):136-152.

Plant, R. E. and L. T. Wilson. 1986. A computer-based pest management aid for San Joaquin Valley cotton. In *Proc. Beltwide Cotton Prod. Res. Conf., National Cotton Council of Amer.*, 22-24.

Roach, J. W., R. S. Virkar, M. J. Weaver and C. R. Drake. 1986. POMME: An expert system to manage pests in apple orchards. Abstracts of papers, 152nd Nat. Meeting, AAAS, Philadelphia, PA.

Robinson, V. B., A. U. Frank and H. A. Karimi. 1987. Expert systems for geographic information systems in resource management. *AI Applications in Natural Resource Management* 1(1):47-57.

Starfield, A. M. and A. L. Bleloch. 1986. *Building Models for Conservation and Wildlife Management.* New York: Macmillan Publishing Company.

Stefik, M. and D. G. Bobrow. 1986. Object-oriented programming: Themes and variations. *AI Magazine* 6:40-62.

Stone, N. D., R. E. Frisbie and J. W. Richardson. 1987. COTFLEX, a modular expert system that synthesizes biological and economic analyses: The pest management advisor as an example. *Proc. Beltwide Cotton Conf., National Cotton Council of Amer.* (In press).

Stone, N. D. and A. P. Gutierrez. 1986. Pink bollworm control in southwestern desert cotton. I. A field-oriented simulation model. *Hilgardia* 54(9):1-24.

Stone, N. D., R. N. Coulson, R. E. Frisbie and D. K. Loh. 1986. Expert systems in entomology: Three approaches to problem solving. *Bulletin of Entomological Society of America* 32:161-6.

Tanimoto, S. L. 1987. The Elements of Artificial Intelligence. Rockville, MD: Computer Science Press.

White, W. B. and B. W. Morse. 1987. Aspenex: An expert system interface to a geographic information system for Aspen management. *AI Applications in Natural Resource Management* 1(2):49-53.

Winston, P. H. 1984. *Artificial Intelligence,* 2nd Ed. Menlo Park: Addison-Wesley Publishing Co.

Woolley, J. B. and N. D. Stone. 1987. Application of artificial intelligence to systematics: SYSTEX—A prototype expert system for species identification. *Systems in Zoology* 36(3):248-267.

5

Computer Development of Insect Population Models

Ivan L. Berry and Ronald E. Stinner

Introduction

This chapter describes methods for numerically simulating the fundamental variables of population size, age, physiological development, mortality, reproduction, and movement in single-insect population models. The relationships among these quantities were defined long before the advent of the digital computer but could be used only for idealized situations. Digital computers reduced the solution of complex sets of differential or difference equations to a comparatively simple process of organizing and repeatedly modifying arrays of numerical values. The major and complex effects of numerous environmental factors on each population component can be readily simulated in digital computer models. Previously, environmental effects usually had to be assumed constant before population models could be formulated and solved. This chapter is concerned with numerical algorithms that allow environmental effects to be simulated, rather than with descriptions of those effects.

Useful models have been developed without consideration of the principal population parameters by empirically relating population size to various environmental factors. However, by omission of well-established relationships, such models lack generality and may fail to provide important information about particular parts of populations.

The relationships among population parameters are so fundamental to population dynamics that they help form the basic internal structure of many modern population models, and certain general results of simulations can be predicted from a knowledge of the earlier analytical solutions. Although modern population models avoid the problem of analytically formulating and solving for the relationships among key population parameters, some of the classical models of population growth will be discussed here.

Classical Birth and Death Models

Several deterministic models that describe birth, death, and the interactions of different species are based on a family of differential equations presented by Lotka (1925). These equations relate the time derivative of population size to various functions of population size and have been widely studied and discussed in demographic and ecological literature (Bailey, 1964; Pielou, 1969). The variation of environmental factors and the effects of age on other population parameters are generally ignored. The simplest of these models is the exponential, or Malthusian, equation for population growth:

$$\frac{dN(t)}{dt} = b\,N(t) \tag{5.1}$$

where

N(t) = population size at time t
b = birth rate per individual per unit time

Another common differential equation for population growth is:

$$\frac{dN(t)}{dt} = N(t)\,f(\,N(t)) \tag{5.2}$$

where f(N) is any function restricted so that df(N)/dN is negative. Thereby, the size of any population will ultimately be limited as f(N) approaches zero with increasing time. The simple case of f(N) = a – bN results in the Verhulst-Pearl logistic equation (Verhulst, 1838; Pearl, 1925). The logistic equation for population growth has all of the limitations of the exponential version, except that the constant environment is assumed to be limited in space or food, so that population size approaches a maximum as time increases. The equation has been fitted to experimental data with excellent agreement in certain cases, but any advantage over other sigmoid curves is questionable. In fact, convincing arguments have been made for more complex forms of f(N), such as c(K – N)/(K + eN) (Smith, 1963).

Population limiting by other species whether by competition, predation, or parasitism, is probably more common than the simple limits imposed by the logistic equation or other functions for single species. For competition between two species, the following version of the logistic differential equation has been used (Pielou, 1969):

$$\frac{dN_1}{dt} = N_1\,(a_1 - b_1N_1 - b_{12}N_2)$$

and

$$\frac{dN_2}{dt} = N_2\,(a_2 - b_{21}N_1 - b_2N_2) \tag{5.3}$$

where

N_1, N_2 = sizes of populations for species 1 and 2
a_1, b_1 = logistic parameters for species 1
a_2, b_2 = logistic parameters for species 2
b_{12}, b_{21} = effect of species 1 on 2 and vice versa

Equation 5.3 cannot be implicitly solved but can be analyzed to determine the ultimate fate of the two populations. Depending on the relative magnitude of the coefficients, both populations will reach a stable equilibrium, or one of them will become extinct. The major fault of this relationship is that it fails to allow for the low probability that two species occupy exactly the same habitat or for the even lower probability that they compete identically in the same habitat.

Several other expressions similar to equation 5.3 have been used to describe host-parasite or host-predator relationships. The best-known are the Lotka-Volterra equations (Lotka, 1925):

$$\frac{dH}{dt} = (a_1 - b_1 P)\,H$$

and

$$\frac{dP}{dt} = (-a_2 - b_2 H)\,P \qquad (5.4)$$

where

H, P = size of host and parasite populations, respectively
a_1, a_2 = exponential growth parameters for populations H and P, respectively
b_1, b_2 = effect of predator on host population and vice versa

The solutions for equation 5.4 generally show that H and P oscillate, with P lagging H by about 1/4 of a period. A number of differential equations similar to equations 5.3 or 5.4 have been proposed for describing competitor-parasite-predator relationships, with solutions and analyses becoming progressively more difficult. Two forms described by Leslie and Gower (1960) may be more realistic than equation 5.4 for modeling, because solutions for both parasite and host populations exhibit damped oscillations after perturbation, thereby reducing the probability of extinction.

Several of the above models have been expressed in the form of difference equations which are very useful for describing the population growth of species with discrete generations. An entire class of these difference equations evolved from a model by Nicholson and Bailey (1935):

$$H_{t+1} = H_t\, S\, F P_1\, P_2\, P_3 \ldots \qquad (5.5)$$

where

H_t = population size of *tth* generation,
S = proportion of females
F = natality
P_i = proportion surviving the *ith* stage.

P_i may also represent the proportions of the same or different stages that survive separate, independent causes of mortality. Key factor analysis for interpreting data from field populations is one of the more important applications of this model (Morris, 1959; Varley and Gradwell, 1968). The differential equations for multiple species have also been expressed in the form of difference equations for discrete generations (Royama, 1971; May, 1973).

Deterministic Age-dependent Models

Applications of the above models are greatly limited by their failure to describe the effects of age on reproduction and mortality. Age is a critical factor in describing populations of species with overlapping generations or with limited seasonal synchronization. The rather fundamental relationships of age-dependent mortality and age-dependent reproduction to population growth can be easily expressed in difference and differential equations, but explicit solutions are possible only for idealized cases. Some of the explicit solutions are presented here because modern insect population models, regardless of complexity, are based on numerical solutions of the same fundamental equations.

The relationships of age to mortality and reproduction are often described by means of life-tables, which are tabulations of these factors according to age. Ideally, classifications are by equal increments of age; in practice, experimental data may be only available for developmental stages of unequal duration. The importance of life-tables for the study of animals is indicated by Southwood (1966), who comments that the first nine chapters of his book culminate in the subsequent analysis of life-tables in the last five chapters. Survival alone is the subject of considerable theory and study (Gross and Clark, 1975).

Lotka's Equations. For a deterministic model of population growth within a constant, unlimited environment, let L(x) be the surviving proportion of a cohort of age x, and let m(x) be the rate of reproduction for each survivor of age x. For this situation, Lotka (1925) proved that the distribution of ages within a population converges toward a fixed function or "stable age distribution" that depends entirely on L(x) and m(x). Without any limitation on population growth, the size of a stable population is exponentially related to time, with the following characteristics:

$$\int_0^\infty L(x)\, m(x) \exp(-r\; x)\, d\, x = 1 \tag{5.6}$$

$$1/b = \int_0^{\infty} L(x) \exp(-r\ x)\, d\ x \tag{5.7}$$

$$d = b - r \tag{5.8}$$

$$R_o = \int_o^{\infty} L(x)\, m(x)\, d\ x \tag{5.9}$$

$$T = \frac{\ln(R_o)}{r} \tag{5.10}$$

where

r = the exponential, or intrinsic rate of increase
b = the mean individual birth rate
d = the mean individual death rate
R_o = the net, or generation growth rate
r = the mean duration of a generation

The above characteristics have been widely used in ecological research on insects, with detailed discussion of their meanings and of methods for calculation (Birch, 1948; Cole, 1954).

McKendrick or Von Foerster Equations. McKendrick (1926) and Von Foerster (1959) have both been credited for equations that describe population growth with age-dependent birth and death rates without the assumption of stability. Populations are described at a point in time by a density function of age, n(t, x), which equals the total population size when integrated over all possible values of age x. With L(x) and m(x) having the same definitions as for Lotka's equations:

$$n(t, x) = L(x)\, n(t - x, 0) \tag{5.11}$$

$$n(t, x) = \frac{L(x)\, n(t - x + r, r)}{L(r)} \tag{5.12}$$

$$n(t, x) = \int_x^t L(r)\, m(r)\, n(t - r, x)\, d\ r \tag{5.13}$$

where all arguments are zero or greater. Equations 5.11 and 5.12 are also expressed by the differential equation:

$$\frac{\partial n}{\partial t} + \frac{\partial n}{\partial x} = -\,h(x)\,n(t, x) \tag{5.14}$$

with the boundary value of n(0, x). The quantity h(x) is the individual death rate, or hazard rate, and is related to L(x) by:

$$h(x) = -\,\frac{1}{L(x)}\left(\frac{dL(x)}{dx}\right) \tag{5.15}$$

Various forms of the these continuous equations are used to approximate population changes for discrete time intervals. However, the same relationships may also be expressed in discrete versions.

Discrete Time Models of Age-dependent Birth and Death

Cole (1954) used calculus of finite differences to derive a population growth equation from life-tables when data were available for equal time intervals. His methods, with notation from Berry et al. (1973), are used here. Let N^j_t be the number of individuals in a population of age j at time t. Then, discrete forms of equations 5.11, 5.12, and 5.13 are:

$$N^j_t = L_j N^o_{t-j} \tag{5.16}$$

$$N^j_t = (L_j/L_k)\, N^k_{t-j+k} \tag{5.17}$$

$$N^j_t = \sum_{i=j}^{n} L_i\, m_i\, N^j_{t-i} \tag{5.18}$$

Equations 5.16 and 5.17 give the relationship between numbers in one cohort at different times and equation 5.18 is the general expression for the size of any age group as a function of previous numbers in the same age group (n is the maximum age for reproduction). For convenience, N_t in the following equations refers to newly born members, or eggs, of the population. Equation 5.18 is a homogeneous, linear difference equation with constant coefficients and has the general solution:

$$N_t = \sum_{i=1}^{n} C_i (z_i) t \tag{5.19}$$

where z_i is the *ith* root of the characteristic polynomial:

$$(z)\, n - \sum_{i=1}^{n} L_i\, m_i\, (z)^{n-i} = 0 \tag{5.20}$$

and C_i is determined by the boundary conditions of N_t. Since L_i and m_i are positive, equation 5.20 always has only one positive real root, which is always of greater magnitude than any other root. Equation 5.20 is a discrete version of equation 5.6, and comparison shows that the real root, denoted z_m, is equal to the exponential of the intrinsic rate of increase. The other roots are either negative or complex conjugate pairs and result in terms of equation 5.19 of the form:

$$(R_i)^t \left[A_i \cos(\phi_i) + B_i \sin(\phi_i)\right] \tag{5.21}$$

where R_i and ϕ_i are the amplitude and angular frequency, respectively, of the complex or negative roots, and A_i and B_i are determined by boundary conditions. Other population parameters can be calculated from discrete versions of equations 5.7 through 5.10.

Thus, in a constant, unlimited environment, a population oscillates about a power curve determined by the one real positive root. As the magnitudes of the oscillating terms are less than the magnitude of 2 m, the amplitudes of the oscillations decay in relation to population size. An example of this behavior is shown in figure 5.2 for L_i and m_i values from colonized horn flies (fig. 5.1). The real, or dominant, root was 1.23 and the complex pair of roots had a magnitude of 1.19 and a period of 13.8 days.

Equations 5.16, 5.17, and 5.18 can be represented in matrix notation (Leslie, 1945, 1948), where $[N_t]$ is a vector representing all age classes at time t:

$$[N_t] = \begin{vmatrix} N_t^o \\ N_t^1 \\ '' \\ '' \\ N_t^{n-1} \end{vmatrix} \tag{5.22}$$

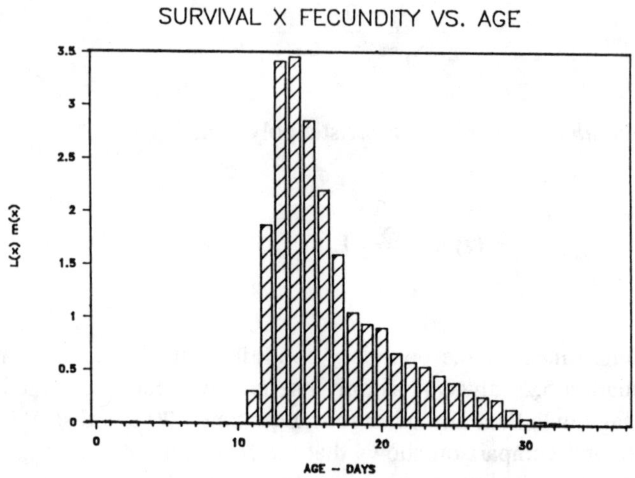

Figure 5.1–Product of survival and natality for colonized horn flies. (Berry et al., 1973.)

Then, the population at one time unit later is:

$$
[N_{t+1}] = \begin{vmatrix} L_1 m_1 & L_2 m_2/L_1 & -- & & L_n m_n/L_{n-1} \\ L_1 & 0 & -- & 0 & 0 \\ 0 & L_2/L_1 & -- & 0 & 0 \\ ---------- & ---------- & -- & ------------ & --------- \\ 0 & 0 & -- & L_{n-1}/L_{n-2} & 0 \end{vmatrix} [N_t] \tag{5.23}
$$

where the matrix of L_i and m_i, designated [A], is the forward projection matrix. If post-reproductive females are counted in the population, values at the right-hand end of the first row may be zero, and the matrix is singular. Equation 5.23 is still valid, but the matrix is assumed to include only reproductive females in the following relationships.

If the age distribution becomes stable in time, then the matrix relationship becomes:

$$
[A][N_t] = z[N_t] \tag{5.24}
$$

where z is a scalar quantity. By standard methods of matrix algebra, the scalar values of z for which equation 5.24 is true can be determined to be the roots of a polynomial in z, which is exactly the same as equation 5.20. Thus, this formulation yields exactly the same results as the difference equation 5.18.

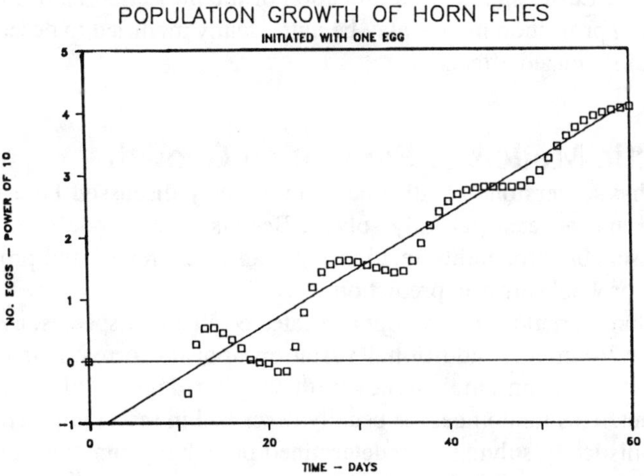

Figure 5.2–Predicted population growth for survival and natality given in figure 5.1. (Berry et al., 1973.)

Leslie (1945) and Pielou (1969) obtained slightly different formulas because they used transitional, instead of cumulative, mortalities and assumed that reproduction occurred before mortality for each time increment. The correct choice of formulation depends entirely on the form of the experimental data.

The projection matrix, [A] has been modified to simulate population limiting by density dependence (Leslie, 1959). For this situation, the projection matrix is a function of time and is defined by:

$$[A]_t = [A][Q]^{-1} \quad (5.25)$$

where [Q] is a square matrix, with each diagonal element i defined by:

$$q_{it} = 1 + aN^{i}_{t-i-1} + bN^{i}_{t} \quad (5.26)$$

Leslie showed that a population described by this matrix initially exhibits damped oscillations as it approaches a stationary level. This population level can be described by the projection matrix obtained by dividing each term of [A] by z_m. Other elaborations of matrices for describing population growth include the development of stochastic versions (Pollard, 1966) and adjustments for unequal stage groupings instead of equal age groupings (Lefkovitch, 1965).

Matrices to describe population growth are naturally suited for computer implementation. In contrast to Lotka's equations, discrete time and age increments are used. The calculation of the array $[N_t]$ provides information on all stages at a given time, which may be more desirable than the calculation of numbers in one stage at several times, as provided directly by the difference equation 5.18. Leslie's (1959) modification of the projection matrix is a simple case of how a projection matrix may be numerically modified to describe a wide variety of time-related effects.

Stochastic Models of Population Growth

Stochastic versions of all models previously discussed here have been formulated and at least partially solved. Because such models are relatively simple, reasonable probability relationships can be expressed and provide some estimate of probable error in predictions.

Random variation in development rates occur in all species, even in very uniform environments; additional variation appears to result from random selection of environmental niches with varying degrees of suitability for development. This randomness is usually described in insect population models by deterministically solving a predetermined probability function for means of developmental time or developmental rate.

The binomial and related discrete distributions are fundamental to population dynamics, simply because the individuals in a single subclass might each be assumed to have the same probability of survival through a given increment of time. Other more complex distributions result from summing individuals from subclasses with different probabilities.

The Poisson process is one of the more powerful stochastic methods for describing population growth. Closed expressions for the probabilities, means, and variances of populations may even be obtained when birth, death, and immigration rates are expressed as functions of time, so that the effects of variable environments may be simulated (Bailey, 1964). One of the greatest shortcomings of the Poisson process is the difficulty of expressing birth and death rates as a functions of age. However, the process may still be useful for describing the survival distributions of individual cohorts by defining death rates as functions of physiological age, or for describing adult populations when death rates are assumed to be independent of age. Bailey (1964) summarizes solutions of the Poisson process from several sources.

The error in most insect population models may be credited more to the failure to adequately describe external environmental effects than to the inability to describe inherent variation in the principle population parameters. Very appropriately, current population models of insects of economic importance seem to emphasize deterministic relationships between the principal population parameters and critical environmental variables. Because of the complexity, probabilistic considerations are usually deferred until completed models can be validated by comparison of simulations with field populations.

Numerical Representation of Population Parameters

Physiological Development of Insects. Because insects are poikilothermic, temperature has a dominant effect on physiological development, which in turn has major effect on the timing and size of populations. The relationship between temperature and physiological development is usually determined by rearing insects at constant temperatures and recording the times required for development through the various growth stages. The oldest and simplest equation to represent these relationships is:

$$\frac{1}{T} = b_0 + b_1 \theta \qquad (5.27)$$

where

θ = average temperature during development

b_o, b_1 = regression coefficients

The above equation leads directly to the concept of degree-days, where the threshold temperature, of zero development, is $-b_o/b_1$, and the time integral (or summation) of temperature above the threshold is equal to $1/b_1$ for full development. Because the inverse of developmental time is usually not linearly related to temperature at both the lower and upper extremes, numerous alternatives to equation 5.27 have been proposed in the literature and used in computer models. Most of these applications have had the common feature that $1/T$ is used to define the rate of development of a given stage at a particular temperature. Then, the integral or the summation of $1/T$ over time from the beginning to an intermediate point of a stage is a relative measure of development, which begins at zero and ends at one if the insect survives. The integral, or summation, is widely used in models to describe the cumulative effects of continuously varying temperatures on development and is designated as physiological age.

As mentioned in the discussion of stochastic models, the developmental times of insects are variable, even when reared in uniform laboratory environments. These variations may be important in describing populations where survival is dependent on part of the immature population being in a narrow range of development during seasonal changes. For *Heliothis* species, normal distributions of developmental time have been used (Hartstack et al., 1976), but distributions over closed intervals may be more appropriate (Stinner et al., 1975). For models of several orchard species, Welch et al. (1978) used the Erlang family of probability density functions, which is a special case of the Gamma distribution. Several researchers indicate that the distributions of developmental time are skewed, with longer tails on the right, and that the standard deviations are approximately proportional to the means. However, the standard deviations for the mobile stages, and sometimes the eggs, may be greater than for the quiescent stages. With several fly species, the skewness is more pronounced, and the minimum developmental time for large samples appears to be more predictable by regression methods than does the mean developmental time (Larsen and Thomsen, 1940; Berry et al., 1977).

Variation in development may be expressed by distributions of developmental rate rather than by distributions of development age. If developmental rate equals the reciprocal of developmental time, each distribution is uniquely determined by the other. Sharpe et al. (1977) used this relationship by assuming that the distribution of average developmental rates was a closed, unskewed parabolic function, which then uniquely determined a skewed distribution for developmental time. Berry et al. (1977) established a probability relationship between development (Q) and physiological age (P) by defining both quantities in terms of minimum total development time (T_{min}) for all immature stages P is the time integral of $1/T_{min}$ for any age, and Q is the minimum value of P for each stage transition. These are related by:

$$y = (P - Q)/Q \tag{5.28}$$

where

P = physiological age in terms of T_{min}

Q = minimum values of P for specified stage transitions

The variable y is assumed to have a Gamma or similar distribution with fixed mean and variance, and the distribution of either P or Q can be determined for fixed values of the other variable. Also, the standard deviation of either variable is proportional to the other variable.

The above methods generally assume that insect development has the same response to temperature for the total span of development. In fact, particular stages or transitions may have different responses that need special treatment in models. Some stages of many insects are capable of thermoregulation (Heinrich, 1981). Developmental rates may also be related to such environmental factors as moisture (Chu, 1987) and nutrition (Sharpe and Hu, 1980), with effects assumed to be proportional at all temperatures in present models. Diapause or other overwintering behaviors will usually involve different temperature responses and be at least partially regulated by photoperiod (Dingle, 1978). Curry et al. (1978) discuss the implications of nonlinear and age-dependent development rates.

If development age for a particular stage transition has the cumulative distribution F (a), then the proportion of individuals with age $a = a_1$ that advances to the next stage by a_2 is given by:

$$R = (F(a_2) - F(a_1)) / (1 - F(a_1)) \tag{5.29}$$

This version of the hazard function (defined in following section) also applies for incremental values of y calculated from equation 5.28.

Mortality. Mortality as a function of age was defined for equation 5.6 as L(x), but experimental data for insect mortality may also be summarized in the form of continuous density or cumulative probability functions for age at death. Another useful function for describing mortality is the hazard rate, which is defined by equation 5.15 and is the individual rate of death at age x, given that the individual has survived to that age. The inverse of equation 5.15 is:

$$L(x) = \exp\left(-\int_0^x h(r)\, d\, r\right) \tag{5.30}$$

The relationships of L(x), the probability density function, and the cumulative probability function to an arbitrary hazard rate are illustrated in figure 5.3. The given hazard function results in a transformed version of the least extreme value distribution. In discrete numerical models, the effects of mortality are conveniently simulated by multiplying the numbers of insects in each subclass of the population at the beginning of each time increment by a survival factor, say S_i, to yield the number of survivors at the next point in time. In the same interval, the effects of several different sources of mortality may each be represented by $S_{i,\,j}$ with:

$$S_i = \prod_{j=1}^{k} s_{i,j} \tag{5.31}$$

The k survival factors in the above expression can be replaced by one minus the dying proportions for the same sources of mortality.

If mortality is expressed in the form of a continuous, cumulative distribution of age at death, G(a), then:

$$S_2 = \frac{(1 - G(a_2))}{(1 - G(a_1))} \tag{5.32}$$

is used to estimate survival factors. Here, 1-G(a) equals L(a) as previously defined. If mortality is expressed as a hazard function, h(a), then:

$$S_2 = \exp\left(-\int_{a1}^{a2} h(a)\, d\, a\right) \tag{5.33}$$

In these relationships, a is generally assumed to be an increasing function of time to represent physiological aging or other biological changes, from birth or any other transition For the special case of h(a) being constant, ages of death are exponentially distributed, as is sometimes assumed for adults.

Insect mortality is often classified as being innate or as being external, but the distinctions are not clear. Conceptually, innate or intrinsic mortality results only from the aging process within an optimum environment, whereas external mortality is caused by various environmental factors, independently of the population parameters. With insects, physiological age is directly affected by

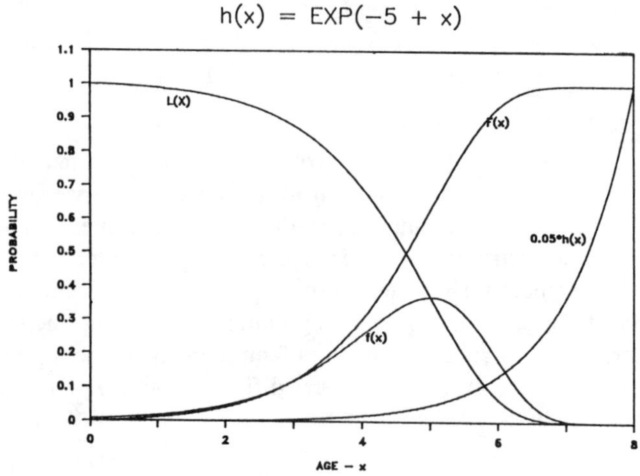

Figure 5.3–Probability functions for survival as a function of age for the specified hazard rate.

environmental temperature, which therefore becomes a factor in determining innate mortality. The temperature relationships of both developmental rates and survival often have the characteristic of being bounded at similar upper and lower temperatures, but the two variables are not necessarily proportional (Casagrande and Haynes, 1976). Also, temperature may interact with many other external variables, such as moisture and nutrition, to affect mortality (Regniere et al., 1981; Chu, 1987). Therefore, innate mortality from laboratory studies may be used as a convenient baseline for simulation of other mortalities but does not fully account for temperature effects. Whether mortality rates are expressed in terms of actual age or of physiological age depends more on the available data and on the type of model than on any inherent relationship with development.

For model development, some mortality may be associated with specific events of short duration and can be described by survival factors that are not derived from time rates. Examples include the transitions between stages, oviposition, and feeding. The timing of such events is critical if mortality is dependent on the environment

The major subject of parasite-predator-prey systems is beyond the scope of this chapter, except to note that survival factors may be used as described above to simulate the immediate effects of parasites and predators. Furthermore, mortality due to parasitism or predation may be related to previous population sizes and thereby provide an indirect form of population limiting. Mortality related directly to density (i.e., intraspecific competition) has been the subject of numerous theoretical papers, but has been rarely included in simulation models (Berryman and Pienaar, 1973; Stinner et al., 1977).

Mortality due to insecticides was described by Jones et al. (1977) by combining the dose-mortality relationship (probit curve) and the exponential decay relationship for insecticide residues. Similar relationships may also be used to describe mortality due to short-term environmental extremes, such as flooding or freezing (Beerwinkle et al., 1978; Williams and Stinner, 1987).

Reproduction. Reproduction is simulated in numerical models by accumulating the expected numbers of offspring from all reproductive substages. In some species, oviposition is adequately described as a continuous rate, which varies according to the degree of development and environmental variables (Regniere et al., 1979). In others, oviposition must be coordinated with particular environmental events, such as a blood meal for mosquitos.

For the adult females of some species, discrete stages of ovarian development can be distinguished by microscopic examination. Development in these stages is related to temperature in the same fashion as in the immature stages, with rates of oviposition being similarly related to temperature. However, these rates may steadily decrease with age, even under optimum conditions, and be very sensitive to environmental variations.

Movement. Several types of insect movement must be described by insect population models. Populations in large areas may be annually reinitiated by migration, sometimes by movement over extreme distances. In other cases, overwintering survivors from relatively few local sites may infest the general area. Intense populations in small areas may be continuously modified by movement among populations. Perhaps the simplest case occurs when populations are uniform over large areas, so that a population may be defined in terms of area, and the numbers entering the population may be assumed equal to those leaving. However, this simplicity is lost with polyphagous species, which may require quantitive description of movement between fields with different crops. In theoretical ecology, insect movement has been studied as part of the searching or foraging behavior of predators and parasites. Research efforts on modeling of agricultural pest movement have been reviewed by Stinner et al. (1983, 1986).

Regardless of the difficulty in quantifying insect movement, recruitment into a population is simply described in models by adding the numbers of incoming insects into the appropriate stages or substages. These numbers would generally not depend on the existing state of the population being modeled, unless recruitment is considered to be the net rate of movement among neighboring populations in "multifield" models (Stinner et al., 1974, 1986). Emigration from a population may be represented as a form of mortality, and could either be proportional to population size or provide a form of population limiting by a nonlinear relationship.

Insect Population Model Structures

In their simplest form, computer models of insect populations are arrays of numbers that represent the sizes of various parts of the population. The separate parts may be distinguished by any of several criteria, including time, age, cohort, and stage of development. In addition to size, the entries may also represent any of the same criteria, plus other measures of physiological status.

Common forms of models and methods for numerically simulation population changes are described in these sections.

Difference Equation Method. The difference equation 5.18 may be numerically implemented very simply by tabulating the numbers of individuals in each cohort at the time of birth. The development of each cohort in time is individually simulated, with appropriate application of migration and mortality factors, to yield the contribution of new births to future cohorts. The effects of the environment are estimated separately for each cohort, so that repeated reference to different environmental conditions over time is generally required. Also, additional calculations are required to describe the composition of the population at a given time. For these reasons, this method has been used only for special cases.

The difference equation method is effectively used for insects with a few discrete generations per year, where the population at any time can be described as a single cohort. Multiple generations may be treated as extra stages of a single generation. For example, Loewer et al. (1974) treated the stages of the summer generations of the European corn borer as additional stages of the overwintering generation. Physiological age was defined by seasonal accumulations of degree-days, with variation in development being described by specifying ranges of accumulated degree-days for each stage transition through the entire season.

Extended Matrix Method. In contrast to the difference equation method, many computer models of insect populations are based on one or more numerical arrays, which describe the entire population at a given time. Numbers within the arrays are then periodically modified to describe the population at future points in time. The process is designated as the extended matrix method here because it is an extension of Leslie's matrix, whereby a population vector is repeatedly multiplied by a forward projection matrix to represent increasing time. In practice, matrix multiplication may not be used, several arrays may be required to adequately describe the current population, and the forward transformation may be a continually changing function of the current environment. This general procedure has been used to describe population growth in a wide range of species (Stinner et al., 1974; Hartstack et al., 1976; Haile and Weidhaas, 1977; Welch et al., 1978; Moon, 1986; Miller, 1986; Wilkerson et al., 1986).

The matrix procedure has the advantages that subtotals for insect numbers, in the various stages can easily be computed for predicting damage levels and effects of current environments can be simulated without repeated reference to prior environments. Also, this structure allow stochastic elements of models to be described as Markov processes, which have an extensive theory and literature.

The population matrix and its transition through discrete time intervals requires that natality, mortality, and migration be expressed as time rates for particular subclasses. The problem of defining exact measures of development for matrix models does not appear to be fully resolved. For many insects, a tabulation of physiological age for each cohort adequately describes the permanent and cumulative effects of previous thermal environments. If development times are assumed to be deterministic, then each cohort retains its distinct identity as it ages and overlapping generations are easily described by tabulating the size and physiological age of each cohort. The problem of describing random variation in development is discussed in subsequent sections.

The effect of temperature on development can be handled as a Markov process because insects with the same degree of development are assumed to be identical, regardless of the variations in previous temperature regimes. In practice, environments during early stages may have other cumulative and permanent effects on survival and natality of the later stages. For instance, natality and mortality of adult female bush flies are affected by environmental moisture levels of the larval stages (Sands and Hughes, 1977; Hughes and Sands, 1979). In matrix models, arrays must generally have an additional dimension for each variable used to describe such cumulative effects with the bushfly, the head widths of emerging female adults were included as a variable to link previous moisture levels to subsequent natality.

Random Distributions of Development. Methods used for describing random variation in developmental times or rates partially determine the internal structure of matrices for describing populations. Because this variation is usually described as a probability distribution of time or of physiological age for a given stage, may modern insect population models divide populations into stages, which are further divided in subclasses according to the time spent within each stage. During each time increment, increasing proportions of the older subclasses are assumed to develop to the point that they are accumulated in the first subclass of the next stage, where they are all necessarily assigned the same physiological age. Movements of living insects between the subclasses and stages are illustrated in figure 5.4, and a more detailed description is given by Wilkerson et al. (1986).

The procedure allows for different distributions of developmental time for each stage. Developmental time can be expressed in the form of different physiological age functions for each stage to accurately describe the cumulative effects of continually varying temperature. The entire population can be described with a minimum of only two vectors (of varying length): one containing sizes of the subclasses and the other containing physiological ages. The subclasses of such models are not true cohorts because they may contain members of different ages. The possibility that different cohorts may remain in distinctly different environments through part or all of their life is ignored.

Another approach is to classify population members according to narrow ranges of development within each stage. Then, an entire population would be represented as a single vector, and the location of an individual within a given cell should completely describe its level of development without reference to physiological age or other cumulative measures. For even more precision, the population could also be classified by age, so that cohorts could be distinguished throughout their life. Movement of insects within the array during time increments would be specified by distributions of incremental development, and these would logically be conditioned by the current levels of

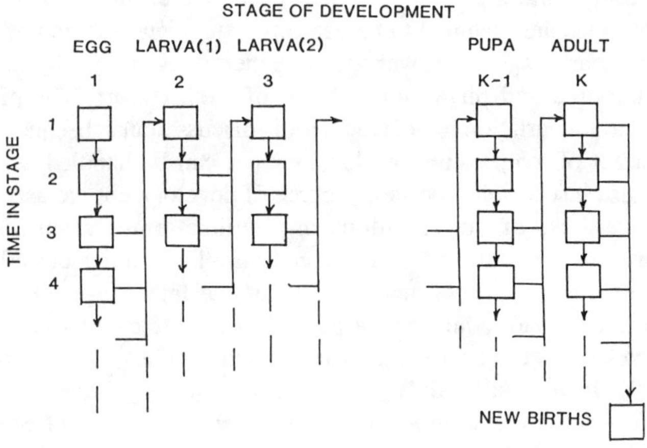

Figure 5.4–Internal structure of a distributed delay model.

development. Development can be mathematically represented by methods discussed in the previous section on developmental rates, but such methods are based on experimental measurements of total stage development times and provide no information about conditional and incremental distributions of development within stages. A partial implementation of a development classification for insect models is provided by the "distributed delay model" (Welch et al., 1978; Manetsch, 1976), with each stage being divided into k equal substages, where k is an integer constant (for all temperatures) of the Erlang distribution and is estimated from an experimentally determined distribution of development times for each stage although never implemented, equation 5.28 provides one means of defining development as a function of physiological age.

Numerical Calculations. If both mortality and development are expressed in terms of cumulative probability functions F(a) and G(a), as used in equations 5.29 and 5.32, the proportion of the initial members in a subclass remaining at t_1 and a_1 would be $(1 - F(t_1,a_1))\,(1 - G(a_1))$. Then at t_2 and a_2, the number remaining in the subclass is:

$$N(t_2) = N(T_1)\,\frac{(1 - F(a_2))\,(1 - G(a_2))}{(1 - F(a_1))\,(1 - G(a_1))} \tag{5.34}$$

The number of members that die is:

$$D(t_2) = N(T_1)\,\frac{(G\,(a_2 - G(a_1))}{(1 - F(a_1))\,(1 - G(a_1))} \tag{5.35}$$

The number of members that advance to the next stage is:

$$R(t_2) = N(T_1)\,\frac{F(a_2) - F(a_1) - F(a_2)\,G(a_2) + F(a_1)\,G(a_1)}{(1 - F(a_1))\,(1 - G(a_1))} \tag{5.36}$$

These expressions assume that F and G have the same argument, which is normally physiological age. However, F may require a different cumulative function of time to describe environmental effects on mortality, and another array would be required to describe the population.

The above equations are simplified if mortality for a single time increment is represented by the single survival factor, S_i. Then:

$$N(t_2) = N(T_1)\, S_2 \frac{1 - F(a_2)}{1 - F(a_1)} \tag{5.37}$$

$$D(t_2) = N(T_1)\,(1 - S_2) \tag{5.38}$$

$$T(t_2) = N(T_1)\, S_2 \frac{F(a_2) - F(a_1)}{1 - F(a_1)} \tag{5.39}$$

Emigration from the population may be represented in these equations as a form of mortality. Immigration may be added to the appropriate subclasses before mortality and advancement to other stages is simulated.

Diapause, adult hibernation, or other periods of dormancy generally require different numerical procedures. Usually, members of only one stage survive periods of dormancy, so that simulations are easily initiated by estimating the numbers of survivors and assuming that they are all in the same cohort. Even without diapause, surviving members of overwintering stages are likely to have similar degrees of development.

Insect Population Model Variations

Numerous variations of the above methods have been used for describing insect populations. One variation is to omit all or part of the relationships among internal parameters and to treat selected measurements as stochastic variables with empirical responses to the environment. Auto- and cross-correlation procedures can be used to partially describe the effects of related species and the effects of reproduction on subsequent generations such tactics are very useful when the relationships between the principal population parameters are generally unknown.

The modeling procedures discussed in this chapter are mostly limited to internal variables of a single population. In practice, microclimates may create several populations of the same species, even on the same plant. Such stratification is described by models for the Banks grass mite (Toole et al., 1984) and for post diapause development in *Heliothis zea pupea* (Logan et al., 1979).

References

Bailey, N. T. J. 1964. *The Elements of Stochastic Processes with Applications to the Natural Sciences*. New York: John Wiley & Sons.

Beerwinkle, K. R., I. L. Berry and S. E. Kunz. 1978. Prediction models for mortality of immature stable flies caused by cold temperatures. Environmental Entomology 7:273-277.

Berry, I. L., J. A. Miller, C. D. Schmidt and J. L. Eschle. 1973. Simulation of horn fly populations. *Transactions of the ASAE* 16(3):482-485.

Berry, I. L., S. E. Kunz and K. W. Foerster. 1977. A dynamic model of the physiological development of stable flies. *Annals of the Entomological Society of America* 70:173-176.

Berryman, A. A. and L. V. Pienaar. 1973. Simulation of intraspecific competition and survival of *Scolytus ventralis* broods (*Coleoptera: Scolytidae*). *Environmental Entomology* 2:447-459.

Birch, L. C. 1948. The intrinsic rate of natural increase of an insect population. *Journal of Animal Ecology* 17:15-26.

Casagrande, R. A. and D. L. Haynes. 1976. A predictive model for cereal leaf beetle mortality from sub-freezing temperatures. *Environmental Entomology* 5:761-769.

Chu, P. C. 1987. Modeling water balance in larval Mexican bean beetles. *Epilachna varivestis* Mulsant. Ph.D. diss., Biomathematics Program, North Carolina State University, Raleigh.

Cole, L. C. 1954 The population consequences of life history phenomena. *Quarterly Review in Biol*ogy 29:103-137.

Curry, G. L., R. M. Feldman and P. J. H. Sharpe. 1978. Foundations of stochastic development. *Journal of Theoretical Biology* 74:397-410.

Dingle, H., ed. 1978. Evolution of insect migration and diapause. In *Proceedings in Life Sciences*. New York: Springer-Verlag.

Gross, A. J. and V. A. Clark. 1975. Survival Distributions: Reliability Applications in the Medical Sciences. New York: John Wiley & Sons.

Haile, D. G. and D. E. Weidhaas. 1977. Computer simulation of mosquito populations (*Anopheles albimanus*) for comparing the effectiveness of control technologies. *Journal of Med. Entomology* 13:553-567.

Hartstack, A. W., J. A. Witz, J. P. Hollingsworth, R. L. Ridgeway and J. D. Lopez. 1976. MOTHZV-2: A computer simulation of *Heliothis zea* and *Heliothis virescens* population dynamics. U. S. Department of Agriculture, Agricultural Research Service, Users Manual ARS-S-127.

Heinrich, B. 1981. *Insect Thermoregulation*. New York: John Wiley & Sons.

Hughes, R. D. and P. Sands. 1979. Modelling bush fly populations. *Journal of Applied Ecology* 16:117-139.

Jones, J. W., E. P. Lloyd and R. E. Stinner. 1977. Insect population mortality: A dynamic model of insecticide effectiveness. *Environmental Entomology* 6:799-803.

Larsen, E. B. and M. Thomsen. 1940. The influence of temperature on the development of some species of diptera. *Vidensk. Medd. Dansk Naturh. Foren.* 8d. 104:1-75.

Lefkovitch, L. P. 1965. The study of population growth in organisms grouped by stages. *Biometrics* 21:1-18.

———. 1945. The use of matrices in certain population mathematics. *Biometrika* 33:183-212.

———. 1948. Some further notes on the use of matrices in population mathematics. *Biometrika* 35:213-245.

Leslie, P. H. 1959. The properties of a certain lag type of population growth and the influence of an external random factor on a number of such populations. *Physiol. Zoology* 32:151-159.

Leslie, P. H. and J. C. Gower. 1960. The properties of a stochastic model for the predator-prey type of interaction between two species. *Biometrika* 47:219-234.

Loewer, O. J., R. T. Huber, J. R. Barrett and R. M. Peart. 1974. Simulation of the effects of weather on an insect population. *Simulation* 22:113-118.

Logan, J. A., R. E. Stinner, R. L. Rabb and J. S. Bacheler. 1979. A descriptive model for predicting spring emergence of *Heliothis zea* populations in North Carolina. *Environmental Entomology* 8:141-146.

Lotka, A. J. 1925. *Elements of Physical Biology*. Baltimore: Williams & Wilkins Co.

Manetsch, T. J. 1976. Time-varying distributed delays and their use in aggregative models of large systems. *IEEE Transactions on Systems, Man, and Cybernetics* SMC-6(8):547-553.

May, R. M. 1973. On relationships among various types of population models. *American Naturalist* 107:46-57.

McKendrick, A. G. 1926. Applications of mathematics to medical problems. *Edinburgh Math Society Proceedings* 44:98-130.

Miller, J. A. 1986. Computer modeling of populations of the horn fly. In *Modeling and Simulation – Tools for Management of Veterinary Pests*, 33-40. U. S. Department of Agriculture. Agricultural Research Service, ARS-467.
Moon, R. D. 1986. Structure and validation of FLYPOP: A simulation model of face fly populations. In *Modeling and Simulation – Tools for Management of Veterinary Pests*, ed. J. A. Miller, 18-27. U. S. Department of Agriculture, Agricultural Research Service, ARS-467.
Morris, R. F. 1959. Single-factor analysis in population dynamics. *Ecology* 40:580-588.
Nicholson, A. J. and V. A. Bailey. 1935. The balance of animal populations. Part 1. *Zoological Society of London* (Pt. 3):551-598.
Pearl, R. 1925. *The Biology of Population Growth*. New York: A. A. Knopf.
Pielou, E. C. 1969. *An Introduction to Mathematical Ecology*. New York: Wiley-Interscience.
Pollard, J. H. 1966. On the use of the direct matrix product in analysis certain stochastic population models. *Biometrika* 53:397-415.
Regniere, J. 1981. *Popillia japonica*: Effect of soil moisture and texture on survival and development of eggs and first instar grubs. *Environmental Entomology* 10:654-660.
Regniere, J., R. L. Rabb and R. E. Stinner. 1979. *Popillia japonica* (*Coleoptera: Scarabaeidae*): A mathematical model of oviposition in heterogeneous agroecosystems. *Canadian Entomology* 111:1271-1280.
Royama, T. 1971. A comparative study of models for predation and parasitism. *Research in Popul. Ecology, Supplement 1.*
Sands, P. and R. D. Hughes. 1977. Mathematical model of the survival rates of adult female bushflies (*Musca vetustissima* Walker) (Diptera: Muscidae) as inferred from field populations. *Bulletin of Entomological Research* 67:675-683.
Sharpe, P. J. H., G. L. Curry, D. W. DeMichele and C. L. Cole. 1977. Distribution model of organism development times. *Journal of Theoretical Biology* 66:21-38.
Sharpe, P. J. H. and L. C. Hu. 1980. Reaction kinetics of nutrition dependent poikilotherm development. *Journal of Theoretical Biology* 82:317-333.
Smith, F. E. 1963. Population dynamics in *Daphnia magma*. *Ecology* 44:651-663.
Southwood, T. R. E. 1966. *Ecological Methods*. London: Methuen & Co.
Stinner, R. E., R. L. Raab and J. R. Bradley. 1974. Population dynamics of *Heliothis zea* (Boddie) and *H. virescens* (f.) in North Carolina: A simulation model. *Environmental Entomology* 3:163-168.
Stinner, R. E., G. D. Butler, J. S. Bacheler and C. Tuttle. 1975. Simulation of temperature-dependent development in population dynamics models. *Canadian Entomology* 107:1167-1174.
Stinner, R. E., J. W. Jones, C. Tuttle and R. E. Caron. 1977. Population mortality and cyclicity as affected by intraspecific competition. *Canadian Entomology* 109:879-890.
Stinner, R. E., C. S. Barfield, J. L. Stimac and L. Dohse. 1983. Dispersal and movement of insect pests. *Annals Rev. Entomology* 28:319-335.
Stinner, R. E., M. Saks and L. Dohse. 1986. Modeling of agricultural pest displacement. In *Insect Flight: Dispersal and Migration*, ed. W. Danathanarayana, 235-241. Berlin: Springer-Verlag.
Toole, J. L., J. M. Norman, T. O. Holtzer and T. M. Perring. 1984. Simulating Banks grass mite (*Acari: retranychidae*) population dynamics as a subsystem of a crop canopy-microenvironment model. *Environmental Entomology* 13:329-337.
Varley, G. C. and G. R. Gradwell. 1968. Population models for the winter moth. In *Insect Abundance*, ed. T. R. E. Southwood, 132-142. Symposium of Royal Entomological Society of London, 4. Blackwell Scientific, Oxford.
Verhulst, P. F. 1838. Notice sur la loi que la population suit dans son accroisernent. *Correspondence Mathematics et. Physics* 10:113-121.
Von Foerster, H. 1959. Some remarks on changing populations. In *The Kinetics of Cellular Proliferation*, ed. F. Stohlman, Jr., 3320-3407. New York: Grune & Stratton.
Welch, S. M., B. A. Croft, J. F. Brunner and M. F. Michels. 1978. PETE: An extension phenology modeling system for management of multi-species pest complex. *Environmental Entomology* 7:487-494.
Wilkerson, G. C., J. W. Mishoe and J. L. Stimac. 1986. Modeling velvetbean caterpillar (*Lepidoptera: Noctuidae*) populations in soybeans. *Environmental Entomology* 15:809-816.
Williams, D. W. and R. E. Stinner. 1987. Respiration and drowning in overwintering pupae of *Heliothis zea* (Boddie). *Canadian Entomology* 119:329-338.

6

Synthesis and Measurement of Temperature for Insect Models

Marshall J. McFarland, Ian R. McCann, and Karen S. Kline

Introduction

Rene Antoine de Reaumur was the first to conduct experiments to determine the relationships between temperature and insect phenology, although he is better known for his work with plants. In work that dates to 1709, he examined the cold resistance and lower fatal limits of caterpillars (spp. unknown) by subjecting them to sub-freezing temperatures (cited by Uvarov, 1931). In 1735 he determined quantitatively the amount of heat necessary to bring a plant from one stage of development to the next stage (cited by Abbe, 1905). He summed the mean daily air temperature between one stage of development and the next and established the relative constancy of this sum from one year to the next. De Reaumur named this the thermal constant.

Abbe (1905) traced the major events in the use of temperature to estimate the phenological development of plants. By the mid-1800s, thermal constants had been determined for nearly 900 species of plants at the Vienna Botanical Gardens. The examination of thermal constants for insects began soon after. Abbe cited work on the determination of thermal constants for the hatching of silkworm eggs and referred to his own work in 1878 on the hatching of grasshopper eggs as a function of temperature. Sanderson (1910) cited this as the first use in America of a thermal constant for insects. Sanderson predicted that such relationships could be put to practical use by farmers.

Apparently the first major quantitative work to relate the entire insect phenology to temperature was done in Texas by Quaintance and Brues (1905). In 1904, they observed 4300 cotton bollworm (*Heliothis zea*) eggs to determine the temperature sums that coincided with hatching. They used the term degree days, although the first use of the term was in 1834 as a measure of the amount of heat required to ripen grain (Abbe, 1905). They used the 43° F (6° C) base developed by Merriam (1894) for the degree day calculations, but found that 45° F (7.2° C) produced a better agreement with the data. The duration of the six larval instars to the nearest hour was observed for 28 lots and a wide range of host plants. The effects of temperature were determined with the use of a commercial chicken egg incubator (26.7 to 32.2° C), an ordinary ice-box

(10.0 to 15.6° C), and a laboratory breeding room with diurnal temperature fluctuations. Additional correlations were found between degree days and depth of burrowing for pupation, pupal stage development, and total life cycle length.

With these developments in 1905, the necessary concepts were in place for the development of insect growth and development models, with temperature as the driver for the models. Specifically, the authors (Abbe, 1905; Quaintance and Brues, 1905) showed relationships between temperature and insect development rate through an entire life cycle, introduced a biological time scale, used both hourly and daily temperatures, used constant temperature environments for the determination of developmental rates, and showed the potential for the practical application of the results.

Other authors were well aware of relationships between insect activity and other weather or climatic elements. For example, Riley (1880) summarized the results of field surveys conducted on the cotton worm (*Aletia agrillacea,* Hubner). In addition to temperature influences on larval development rates, he mentioned the effects of humidity on egg survivability and the effects of a dry, heated soil on the mortality of larvae that dropped to the soil surface for pupation. Fifty years later, Uvarov (1931) based his insects and climate monograph on over 1150 references from eleven languages. Each reference contained one or more relationships between an aspect of insect growth, development, or other activity to an element of weather or climate. Since 1931, several thousand more papers have appeared that describe the effects of single variables such as temperature on a biological response of insects. Each of these insect-weather relationships is a candidate for inclusion in an appropriate model of insect phenology, population dynamics, activity, or other important aspect.

Recent work on effects of weather or climatic elements on insects includes factors other than temperature. Wind is an important factor in insect migration, dispersal, and flight activity (Johnson, 1969; Rainey, 1974). Photoperiod and temperature are determinants of diapause behavior (Omar, 1980; Tauber et al., 1986). Humidity strongly influences mortality of eggs and early larval instars of foliar feeding Lepidoptera species. Soil moisture affects larval activity, pupation, emergence, and mortality of several genera (Hurst, 1975; Henneberry and Clayton, 1983). Radiation influences flight activity, while the absence of radiation from sunlight or moonlight governs nocturnal insect activity (Hartstack et al., 1973). The effects of precipitation are pervasive.

Additional elements were needed to apply the concepts that were developed by 1905. These elements include scouting networks for insect population information, meteorological networks, computerized data management and delivery systems for operational use of the information. Wellington and Trimble (1984) add that a knowledge of the processes, patterns, and scales of biometeorology is needed by entomologists who are engaged in modeling activity.

This chapter addresses the temperature information that may be needed for modeling of the insect environment. Emphasis is placed on temperature, because insect population dynamics are dependent on temperature in most models (e.g., Wagner et al., 1984, 1987; Throne and Eckenrode, 1986). The syntheses of hourly temperatures from daily maximum and minimum temperatures are examined. Requirements for instrument networks are addressed. Descriptions of the microscale and mesoscale environment of the insect are contained in Yao (1981), Hatfield and Thomason (1982), Rosenberg et al. (1983), Wang and Ray (1984), and Wellington and Trimble (1984), so will not be included.

Models for Hourly Temperature Synthesis

The available heat energy in the environment for the growth and development of a poikilothermic organism can be represented by the temperature of that environment (Gutierrez et al., 1980). The summation of temperature has been widely used as a measure of the heat energy available for growth and development of insects and plants. The terminology is imprecise in the literature; degree days, heat units, and thermal units have much the same meaning.

Higley et al. (1986) concluded that the degree day approach is sound for growth rate modeling as long as temperature is recognized as the major determinant of the growth rate and the limitations of the approach are recognized. They list eight factors that influence the adequacy of degree day accumulations in modeling:

- Substrate availability.
- Enzyme availability.
- Approximations and assumptions in laboratory estimates of development.
- Approximations in calculating the developmental minimum.
- Approximations in calculating the developmental maximum.
- Approximations in using single values for developmental thresholds for all stages.
- Thermoregulation.
- Propriety and limitations of temperature data used in the calculations.

Consequently, if the factors are adequately considered, air or soil temperature may be used to determine the rates of insect growth and development in simulation models.

Typically, the time required for an insect to complete a phenological event, such as egg hatching and duration of a stage of growth (or dormancy), is measured under a series of constant temperatures (e.g., see Fargo and Bonjour, 1988; Garcia-Salazar et al., 1988). The time for each stage or event is plotted for each temperature. The development rate is established by plotting the time for each stage or event against the inverse of the temperature summation. The base temperature is obtained by extrapolating the development rate line to the zero development level. The heat units are obtained by summing or integrating temperature over time periods, such as hours or days. The number of heat units is then correlated with the growth or phenological event.

For correlation of heat units with insect growth and development in field environments, the heat units are estimated from the available temperature information. The procedure by Ring and Harris (1983) and Ring et al. (1983) illustrates the selection of the start date of the heat unit summation and the selection of the base temperature. They developed a degree-day model to predict the pupation, adult emergence, and oviposition of the overwintering generation, and hatch and nut entry of the larvae of the first summer generation of the pecan nut casebearer (*Acrobasis nuxvorella,* Nuenzig). They started accumulations on one day increments from 1 January to 30 April for base temperatures between –17.2 to 21.1° C in 0.56° C increments. Comparisons were made between the observed dates of the events and the degree-days for each start date and base temperature. The model variables selected were 12 March and a base of 3.3° C, the combination that produced the lowest root mean square error between observed and model predicted events.

The shape of the function that relates insect growth rate to temperature is generally modeled as sigmoidal. For sigmoidal relationships between growth or development and temperature, no growth or development occurs below a base temperature. Similarly, no growth or development occurs above an upper temperature limit.

If relatively small sections of the curve near the base and upper limit are neglected, the sigmoidal curve may be represented by a straight line (Gutierrez et al., 1980; Harcourt et al., 1983). An underlying assumption for the linear model is that a degree of temperature anywhere in this range represents the same amount of heat energy and produces the same increment of growth and development. Temperatures in excess of the maximum produce development at the maximum rate. With these assumptions, heat energy may be represented by the difference between an average temperature and a base temperature.

The normal curve is also used to describe temperature dependent developmental rates (Higley et al., 1986; Tauber et al., 1986). The curve may either be truncated to the right of the mean or the entire curve may be used, with decreasing development rates for temperatures to the right of the mean. The Logan curve (Logan et al., 1976) accounts for an asymmetry about the maximum temperature, so is of particular utility for the description of development rates in the vicinity of the maximum temperature.

Calculating the heat units is a simple exercise for a constant temperature environment. Similarly, heat units are easily calculated with measurements obtained from automatic data logging equipment for an environment with fluctuating temperatures. The heat units are determined by summing or integrating the temperatures above a base. However, the problem of heat unit calculation is more difficult when the temperature records consist only of daily maximums and minimums. This is typically the case because the only information that is often available is from the climatological observation networks, which record only the daily maximum and minimum temperatures.

The linear method for heat unit calculation from daily maximum and minimum temperatures is the oldest (see Abbe, 1905), simplest and is still used despite its long-recognized inadequacies (Yao, 1981). In the linear method, degree days are calculated as the positive difference between the average daily temperature and a base temperature:

$$DD = \frac{(Tx + Tn)}{2} - Tb \tag{6.1}$$

where DD is the degree day accumulation for a day with maximum, Tx, minimum, Tn, and base, Tb, temperatures.

The mean temperature has to exceed the base temperature. Several variations include different methods of handling base temperatures and incorporating an upper limit. Researchers who have used the linear degree day method include Butler and Scott (1976), Harcourt et al. (1983), Stinner et al. (1973), and Baxendale et al. (1984).

The linear method produces degree day sums that are highly correlated with, but not identical to, sums obtained from actual hourly temperatures or from constant temperature environments. A bias will be introduced because the mean temperature, calculated from daily extremes, differs from the average temperature. The daily average temperature will equal the mean when the diurnal temperature curve is a sinusoidal wave with maximum and minimum values at 12 h intervals.

The accuracy of the degree day estimation may be improved with the use of functions fitted through the temperature extremes. The curve fitting methods are best suited for weather conditions without major day-to-day temperature changes. None of these methods will represent the actual temperature change, however, during frontal passages or other conditions that result in rapidly changing temperatures. Reicosky et al. (1989) noted that several hourly temperature models performed well in clear days in Minnesota. On cloudy days, the model performance was significantly worse. An implicit assumption with the use of curve fitting methods is that the minimum temperature occurs at or near sunrise and the maximum temperature occurs in the late afternoon hours.

Curve fitting methods are suited for computer calculation, input into insect simulation models, and graphical output. Sevacherian et al. (1977) used a triangular method proposed by Lindsey and Newman (1956) for degree-day accumulation. A set of equations was developed for the calculation of degree-days from the triangles formed by connecting successive temperature maximums and minimums with straight lines. The equations were solved one-half day at a time for assumed equal-length days and nights.

Arnold (1960) introduced a sine function that has been modified several times (e.g., Baskerville and Emin, 1969; Allen, 1976; and Snyder 1985, who developed a version for a hand calculator). Higley et al. (1986) presented FORTRAN programs and compared degree days calculated with the sine and the triangular functions. Allsopp and Butler (1987) compared degree days calculated with the sine and triangular functions from soil temperatures. A sine curve is fitted through the maximum and minimum temperatures, one day at a time. The degree days are then calculated by summing the hourly temperatures derived from the sine curve. The temperature at time, t, is:

$$T(t) = \frac{(Tx + Tn)}{2} + \frac{(Tx - Tn)}{2} \times \sin[f(t)] \qquad (6.2)$$

where Tx represents the daily maximum temperature and Tn is the daily minimum temperature.

The diurnal variation of temperature, f(t), is:

$$f(t) = 2 \times \pi \times \frac{(t - 6)}{24} \qquad (6.3)$$

where t is the time in hours from 00 to 24.

With this form, the term (t – 6) sets the time of the minimum and maximum temperatures 12 h apart at 06 and 18.

Diurnal temperature curves are more accurately represented by two or more curves because the heating and cooling curves do not have the same shape. With two or more curves, the actual times of the occurrence of the temperature extremes may be taken into account. Parton and Logan (1981) described the daytime temperature with a truncated sine curve and the nighttime temperature with an exponential curve. Their equations are:

$$T(t) = Tn + \frac{(Tx - Tn)}{2} \times \sin\left\{\frac{Pi(t - tr - b}{d + 2(a - b)}\right\} \text{ for day hours} \qquad (6.4)$$

$$T(t) = Tn(+) + [Ts - Tn(+)] \times \exp\left[\frac{-c(t - ts)}{n + b}\right] \text{ for night hours} \qquad (6.5)$$

where

tr = time of sunrise
b = the lag in hours between sunrise and the minimum
a = the lag in hours between noon and the maximum
d = day length in hours
c = a shape parameter for cooling
n = night length in hours
Ts = sunset temperature at time ts
Tn(+)= the minimum the next day

In equation 6.5, t is, for example, 18 at sunset, 24 at midnight, and 30 at sunrise. Times of sunrise and sunset are (ASHRAE, 1982):

$$\sin(al) = \cos(L) \times \cos(del) \times \cos(H) + \sin(L) \times \sin(del) \qquad (6.6)$$

where

al = solar altitude from horizon
L = latitude
del = solar declination angle
H = the hour angle from solar noon (15°/h)

The solar declination angle, del, is:

$$del = 23.45 \times \sin\left(\frac{2 \times Pi(284 + CD)}{365}\right) \qquad (6.7)$$

where CD represents calendar day.

At sunrise and sunset, the solar altitude, al, is zero. True sunrise and sunset (the top of the sun at the horizon), including atmospheric refraction, are given in the Smithsonian Tables (List, 1949). The differences of less than one degree can be ignored for calculations of times of sunrise and sunset in the curve fitting models.

Rearrangement of equation 6.7 produces:

$$\cos(H) = \frac{-\sin(L) \times \sin(del)}{\cos(L) \times \cos(del)} \qquad (6.8)$$

$$H = \arccos\left[-\tan(L) \times \tan(del)\right] \qquad (6.9)$$

Sunrise and sunset times are local solar noon minus or plus H/15, respectively.

Wann et al. (1985) compared the sine-exponential model of Parton and Logan (1981) with two sine-sine models that did not incorporate sunrise and sunset. For five locations in North Carolina for periods up to nine years, they concluded that the sine-exponential was a better fit to the actual hourly temperatures. In the comparison by Wann et al., the accuracy of the sine-exponential model was not substantially decreased with an averaged set of parameters (a, b, and c) for the five North Carolina locations. Their parameter values were:

a = 2.59 (the lag between noon and time of maximum)
b = 1. 55 (the lag between sunrise and time of minimum)
c = 2.20 (the shape parameter for the exponential cooling curve)

Parton and Logan reported values for Denver as a = 1.80, b = 0.88, and c = 2.20.

Eckersten (1986) modified Parton and Logan's sine-exponential model to a sine-sine-exponential model by adding an equation for the temperature decrease from the maximum to sunset. Eckersten's equations are:

$$T(t) = Tn + (Tx - Tn) \times \sin\left\{\frac{Pi(t - tr)}{d(1 + 2a')}\right\}$$

$$\text{for } tr \leq t < tx \tag{6.10}$$

$$T(t) = Tn(+) + [Tx - Tn(+)] \times \sin\left\{\frac{Pi(t - tr)}{(d)(1 + 2a')}\right\}$$

$$\text{for } tx \leq t \leq ts \tag{6.11}$$

$$T(t) = Tn(+) + [Ts - Tn(+)] \times \exp\left[\frac{-c(t - tr - d)}{n}\right]$$

$$\text{for } ts < t < tr(+) \tag{6.12}$$

where

a′ = (tx – 12)/d
tx = time of the maximum
tr(+) = time of sunrise the next day

All other terms are as previously defined. Ts is determined from equation 6.11.

McCann (1985) developed a sine-sine-sine model with a truncated sine term to describe nocturnal cooling. His form is:

$$T(t) = Ts + [Ts - Tn(+)] \times \sin\left\{\frac{Pi(t - ts + 2n)}{2n}\right\} \tag{6.13}$$

where all terms are as previously defined. The sine curve produced a better fit of nocturnal cooling curves than an exponential curve in summer months for College Station, Texas. To account for a slower rise of temperature than described by the models of Parton and Logan (1981) and Eckersten (1986), McCann (1985) used a one-quarter of a sine wave instead of one-eighth. His sine curve for increasing temperature is:

$$T(t) = \frac{(Tx + Tn)}{2} + \frac{(Tx - Tn)}{2} \times \sin\left\{\frac{Pi(t - tr)}{tx - tr} - \frac{Pi}{2}\right\} \tag{6.14}$$

where all terms are as previously defined. In this portion of the sine wave, inclusion of the b parameter may not be needed.

The values of the a, b, and c parameters in the sine-exponential, sine-sine-sine, and the sine-sine-exponential models should be established for the region and season of interest. Local climatology influences these values. The maximum temperature will occur closer to local noon in an arid, high elevation environment than in a humid, low elevation environment. The lag between sunrise and the time of the minimum will also be less for an arid, high elevation environment than an environment with increased atmospheric absorption and scattering of solar radiation.

Taylor (1981) fitted a Fourier series through the maximum and minimum temperatures for an entire year for simulation of insect development rates for numerous species along latitudinal and longitudinal gradients. The Fourier series had two components; one for the seasonal variation of temperature and the other for the diurnal variation. The form was:

$$T(t) = Tm - \left(\frac{a2}{2}\right) \times \cos\left(\frac{2 \times Pi \times t}{365} - \frac{2 \times Pi \times tp}{365}\right) - \left(\frac{a1}{2}\right) \times \cos(2 \times Pi \times t) \quad (6.15)$$

where

Tm	=	the annual mean temperature
a1 and a2	=	amplitudes
t	=	time (days)
tp	=	the seasonal lag of temperature that is a function of both the latitude and elevation

Reichenbach and Stairs (1984) also used a Fourier series through a long period of record for temperatures. In their work, the temperatures to describe varying life spans of the western spruce budworm [*Choristoneura occidentalis*, Freeman (*Lepidoptera: Tortricidae*)] were then calculated analytically from the Fourier series. A more rigorous use of Fourier series was used by Russell (1984). The individual terms of the Fourier series may be integrated or differentiated without the need to separately examine each term. However, the basic assumptions associated with temperature analysis with Fourier series are much the same as the assumptions for the sine method.

A cubic spline curve may also be fitted through daily maximum and minimum temperatures when degree days from a climatological record is desired. The cubic spline is a third order polynomial that passes through the data points. Descriptions of the cubic spline are in most numerical analysis text and reference books.

Table 6.1. Root mean square errors (C) of four methods for calculation of hourly temperatures

Location	Eckersten	McCann	Sine	Cubic Spline
Houston, TX July 1973 (a = 2.10)	1.55	1.67	2.77	1.98
El Paso, TX July 1973 (a = 2.61)	1.98	2.06	4.54	2.40

Comparison of Models

Hourly temperatures for July 1973 from Houston and El Paso, Texas, were used to compare the Eckersten, McCann, cubic spline, and sine models with actual hourly temperatures. The climate of Houston in July is characterized by humid flow from the Gulf of Mexico. The climate of El Paso is an arid environment, with low humidity and relatively clear skies. The root mean square errors (RMSE) for the 724 hourly temperatures generated by the four methods compared with the observed temperatures are presented in Table 6.1. The Eckersten and McCann models produced similar results; both were more accurate than the sine and cubic spline models.

Because an insect simulation model may be more sensitive to degree-hour or degree-day accumulations than to hourly temperatures, degree-days were also compared for the models. The results are presented in Table 6.2. It was interesting to note that the degree-day accumulations were similar for all models for base temperatures of 8 and 20° C. The models tended to overpredict the degree-days.

The McCann model and the Eckersten models were compared for Kansas City, Missouri; Tallahassee, Florida; Madison, Wisconsin; Amarillo, Texas; Pocatello, Idaho; and Bishop, California. These sites represented a range of environments from humid to arid, with diurnal temperature ranges from 11° C at Kansas City and Tallahassee to 23° C at Bishop. June 1983 and July 1986 temperatures from the Local Climatological Data (data at three-hour intervals) were used for the comparison of the models. The McCann model was more accurate with humid environments with a diurnal temperature range less than 14° C and the Eckersten model was more accurate in semi-arid environments with a greater diurnal temperature range.

Time of Observation

In climatological records, such as the Climatological Data published by the National Climatic Data Center (NOAA), Asheville, North Carolina, the reported minimum and maximum temperatures are for the preceding 24-hour period. When the reporting time is midnight (for relatively few federal stations in the climatological network), the extremes always occur on that day. When

Table 6.2. Monthly totals of accumulated degree-days of the observed temperature data and the four methods used to estimate hourly temperatures
[Base temperatures (T_b) of 8° C and 20° C were used]

Location	Observed	Eckersten	McCann	Sine	Cubic
Houston, TX					
July 1973					
T_b = 8° C	604	629	608	618	617
T_b = 20° C	240	265	243	253	253
El Paso, TX					
July 1973					
T_b = 8° C	556	582	551	561	559
T_b = 20° C	198	219	194	204	202

the observation time is in the morning, normally 0600 to 0800, the reported maximum temperature normally occurred the previous day. The minimum temperature reported may have occurred the previous day also in the case of a warming trend. When the observation time is in the afternoon, normally 1700 to 1900, the minimum normally occurred on the day of the report, but the maximum may have occurred on the previous day. If stations with differing reporting times are included in the same data set, a bias is introduced. Dale (1981) reported differences in degree day accumulations of six percent as a result of time of observation. He also reported that conversion of a station with an afternoon reporting time to a morning time introduced a bias of about –1° C. Others who noted biases of about 1° C as a result of observation time include Rumbaugh (1934) and Mitchell (1958).

Measurement Sites

The variation of the important environmental variables in the plant ecosystem has a major impact on the selection of the measurement sites for a data collection system. The measurement of temperature of the insect environment provides a good example. An insect in contact with leaf will have the same thermal environment of the leaf, modified slightly by the distance its body is from the leaf and the direct or indirect radiation load on the insect. The energy budget of the leaf is controlled by the energy transfer mechanisms of radiation, convection, and transpiration. The canopy is warmed by solar radiation and sensible heat advection and is cooled by thermal radiation and transpiration. These processes are spatially uniform except for sensible heat advection. Brakke et al. (1978) and Rosenberg and Verma (1978) discuss the effects of heat advection from a dryland area into irrigated fields. The sensible heat advection in the form of higher air temperatures was sufficient to significantly increase the daily crop transpiration near the edge of the irrigated field. Burman et al. (1975) made ground level measurements along a 50 km transect from dry sagebrush desert into the center of the irrigated area in the Snake River Plain of Southern Idaho. From the measurement sites in the middle of irrigated alfalfa fields, they showed that the irrigated area was several degrees C cooler, the wind speed was reduced 40%, and the vapor pressure deficit decreased several mb.

Within the field, the mid-day temperature difference between fully watered and water stressed canopies may reach 10° C in an arid environment (Jackson, 1982). Higher air and canopy temperatures at the edge of a field may result in moisture stressed canopies. These edge effects (plants with moisture stress) are frequently important as the initial infestation sites for insects and mites. These edge effects at the scale of the individual field or of the region indicate that weather measurement sites be selected as close as possible to the site of the insect habitat. Even with the most representative measurement sites, there will still be variations of temperature throughout the insect habitat.

Allen (1988) characterizes this problem as,

> ". . . one may have a developmental rate function of temperature, but be frustrated because the organism is in a distribution of temperatures that is poorly known. We usually take an average temperature, plug it into the function, and hope that everything will work out . . . It is rarely considered that the problem might be in how the average of the function is estimated when its variable has a distribution."

He presented a second-order approximation method that is independent of the statistical distribution of the environmental variable, such as temperature. The mean and variance of the variable in the environment have to be known, however, to find the average value of the nonlinear function. With the aid of appropriate sensors, multiplexers, and automatic data loggers, this task is not difficult to perform on an aperiodic basis. The relationship of the mean and variance of the insect environment to the measurement point of the environmental variables should ideally be established and aperiodically verified during important phenological events.

In a field situation, two types of measurement sites may be advisable. Base weather stations that are part of a large area network, such as an agricultural weather network, would have standardized placement and exposure of sensors. Standard climatological measurements are independent of the underlying surface (which ideally is grass). For example, solar radiation is a climatological variable that is independent of the underlying surface; net radiation is a function of the albedo and thermal radiative characteristics of the underlying surface. Similarly, the temperature within the crop canopy will be strongly influenced by these factors, in addition to advective gradients across the field. A second type of weather station is a non-permanent site at the most representative location for the specific insect(s) of interest. The relationship between the environmental variables at the two types of weather stations should be established, perhaps with Allen's (1988) technique.

The establishment of specific insect habitat and standard weather measurement sites will facilitate the incorporation of weather data from a number of sources into the data base for the models. The standard weather measurement sites will also facilitate operation of multiple insect habitat models from a standard measurement network.

Measurement Networks

The instrumentation for a weather measurement station is described in sources such as Fritschen and Gay (1979), Wang and Ray (1984), and Rosenberg et al. (1983). A weather measurement station for insect growth and development model data should be based on an automatic data logger. Features of automatic data loggers include:

- A modular system with a number of input channels, both analog and digital.
- Suitability for virtually all types of electronic environmental sensors. Voltage from a power supply for excitation voltage may be monitored on an input channel.
- Programmable sampling rate and signal processing. This normally consists of conversion of output in millivolts to appropriate units, selection of maximums, minimums, totals; selection of instantaneous or time-averaged readings, and selection of output device(s).
- Internal memory storage with on-line or interrogated transfer to cassette tape, printer, modem, or computer file.
- Battery power capability.
- Capability for addition of a multiplexer to increase capacity for similar sensors. For example, up to 32 thermocouples may be connected to a single input channel with the use of a multiplexer.

For continuous monitoring or operational based pest management program (e.g., Croft et al., 1976; Haynes et al., 1973), the automatic data loggers may be equipped with a telephone dial-up modem, hard wire, or telemetry systems.

For regional validation and calibration of models or for real-time operation of insect models, a network of stations is required. The design of this network is a function of:

- The size, shape, topography, and location of the area.
- The variation of the environmental variables within this area. The variables of major importance or sensitivity, such as air temperature, will take precedence.
- The number of weather stations that will comprise the network. This a function of the density of existing observations, the initial cost of the system additions, the time and cost for system (including sensors), maintenance, and the resources available for system operation.
- The availability and accessibility of representative sites in the fields of cooperating producers or agencies.

Network design is a process of optimizing cost and precision of measurement. If cost were not a factor, an over-abundance of measurement sites could be selected. The high density of sampling sites for the variables would reduce uncertainty to inconsequential levels. Cost is an unavoidable factor, so a balance must be obtained between the cost of the system and the resultant uncertainty in the estimation of the spatial distributions of the variables from a limited number of sampling points. The allowable uncertainty is quantifiable from the sensitivity of the models. This may be expressed in a number of ways, from the parameterizations of the most sensitive variable to the definition of an acceptable error in the predictions of the model. The relationships between the cost of the data system and the value of the data collected are not obvious.

As a general statement, the marginal benefits produced by additional data should never be less than the cost to collect the data. In the absence of this framework, two basic principles of network design are suggested. A network operated on a fixed budget should be designed to minimize the error of estimation of the primary or most sensitive variables. A network operated on a criteria of minimum acceptable error of estimation should be designed for a minimum cost (Rodriguez-Iturbe and Mejia, 1974).

The approach to the design of the network may be directed by the predominance of any one of four major factors of area characteristics, variation of variables, number and cost of stations, and availability of sites. The process of network design (and redesign) may become a series of ad hoc responses to changes in these four factors. Typical questions that prompt design or redesign of a network are: If only five data loggers are available, where should they be located? If the area of the program is expanded, how many new weather stations are required? If variations in thermal units from one site to the next are deemed significant, how should the network be redesigned? If funds become available for an additional weather station, how should the network be redesigned? Or, should the new weather station be placed in the largest gap, either spatially or variability in coverage?

Several options are available to enhance the network design. The first step to each option would be to inventory the available climatological information for the area. If data is available from several locations within the larger area or region, a structure function analysis may be useful. The correlation coefficients are calculated for temperature (or other weather variable of interest) for each pair of stations. The correlation coefficients are plotted as a function of distance. The resulting function is known as a structure function. The information obtained from a structure function analysis is the variability of the weather variable with distance. The shape of the structure function is normally an exponential decay from an auto-correlation coefficient of 1.0. If the shape of the function is not apparent, distance is not the major determination of the variability. For example elevation effects or other locational features may have a greater effect on the value of the environmental variable. The degree of isotropy of the function may also be determined from structure function analysis. An isotropic field will have a structure function that is not a function of direction (e.g., plot east-west pairs of stations separately from north-south pairs of stations).

Another technique involving the correlation coefficient is to determine the correlation coefficients for a weather variable for all stations with a single station (the key station). These correlations coefficients are plotted on a scale map of the weather station network. A visual analysis of the contours of equal correlation coefficient will show the degree of anisotropy of the pattern. Anisotropy will be produced by effects of latitude, elevation, proximity to large water bodies or lands of differing cultural practices, and prevailing wind direction. For an isotropic pattern, a measurement network should have stations with equidistant spacing. However, if the network has a pronounced degree of anisotropy, the spacing of the stations should be closest in the direction of the greatest gradient of the contours. A methodology to test in terms of the sensitivity of the model is to select a correlation coefficient, such as 0.9 for air temperatures, as the criteria for spacing. The model is operated with a temperatures of both sites and the results are compared. If the differences are acceptable in terms of model sensitivity, then a satisfactory spacing will be obtained with the use of 0.9 as a minimum acceptable correlation coefficient between the time series of stations in the network. If the differences are not acceptable, the value of the correlation coefficient as the basis for station spacing should be increased.

Fujioka and Fosberg (1981) proposed a triangulation method for allocation of weather station points that represent an advantage over a traditional method of placing new stations in the largest gap in coverage. An assumption is made that the difficulty of describing a pattern from a number of sample locations is related to the variability in the pattern, as measured by the variance at each point. In their scheme, a system of triangles is constructed with existing measurement points at the vertices. The variance of the weather variables is calculated at each vertex. The location of a specific number of new stations to be added to the network is determined from a ratio of the products of the variances to the areas of the triangles.

A method for selecting measurement network sites based solely on similarities in the measurements at existing points is given by Brunner et al. (1985). They determined from a cluster analysis the number and location of temperature measurement sites to adequately represent the tree fruit production sites in the Wenatchee River Valley in Washington. Degree-day accumulations through the season were calculated from a network of 31 sites in 1979 and 36 sites in 1980. The degree-day accumulations for each station were plotted on the same graph on a monthly basis. Clusters of stations that accumulated degree-days at nearly the same rate were evident. Eleven sites were required to provide an accurate resolution of temperature for degree-day calculations. Some sites separated by a considerable distance were clustered, while some adjacent sites were characterized by large differences in degree-day accumulations. Multiple regression on elevation, slope aspect, and slope degree did not provide an adequate explanation of the temperature variability between sites. This study by Brunner et al. illustrates an approach commonly noted in the design of networks for environmental monitoring. An initial network of greater density than required is necessary to define the number and location of measurement points that will constitute the final network.

Summary

Ideally, the source of air temperatures for development, validation, calibration, or operation of insect growth and development models would be from a temperature sensor located in the immediate environment of the insect that is modeled. The data from the sensor would be recorded at intervals as short as 10 to 15 minutes. The temperatures would then be processed into a computer database for ease of modeling. Often, however, daily maximum and minimum temperatures are the only source of temperature information for a model that requires updating on intervals of an hour or less. Various curve fitting techniques are suitable for synthesis of hourly (or more frequent) temperatures for situations without air mass changes. Design of environmental instrument networks for insect modeling is frequently hampered by lack of resources. Several techniques are available for the selection of the optimal number of weather stations as a function of model sensitivity and resource availability. The simplest is with the analysis of correlation coefficients between pairs of stations in an existing network, but other techniques are available.

References

Abbe, C. 1905. The relations between climate and crops. Weather Bureau Bulletin No. 36. Washington, DC.

Allen, J. C. 1988. Averaging functions in a variable environment: A second-order approximation method. *Environmental Entomology* 17(4):621-625.

Allen, J. C. 1976. A modified sine wave for calculating degree days. *Environmental Entomology* 5(3):388-396.

Allsopp, P. G. and D. G. Butler. 1987. Estimating day-degrees from daily maximum-minimum temperatures: A comparison of techniques for a soil-dwelling insect. *Agriculture and Forest Meteorology* 41:165-172.

Arnold, C. Y. 1960. Maximum-minimum temperatures as a basis for computing heat units. *Proceedings of the Society of American Horticultural Science* 76:682-692.

ASHRAE. 1982. Solar energy utilization. In *1982 Applications Handbook,* 57.1-57.7. Atlanta GA: Amer. Soc. of Heat., Refrig., and Air Cond. Engineers.

Baskerville, G. L. and P. Emin. 1969. Rapid estimation of heat accumulation from maximum and minimum temperatures. *Ecology* 50(3):513-517.

Baxendale, F. P., G. L. Teetes, P. J. H. Sharpe and H. Wu. 1984. Temperature-dependent model for development of nondiapausing sorghum midge (*Diptera: Cecidomyiidae*). *Environmental Entomology* 13(6):1572-1576.

Brakke, T. W., S. B. Verma and N. J. Rosenberg. 1978. Local and regional components of sensible heat advection. *Journal of Applied Meteorology* 17:955-963.

Brunner, J. F., S. C. Hoyt and G. V. Tangren. 1985. Temperature monitoring sites for tree fruit pest management in north central Washington. *Agriculture and Forest Meteorology* 33:315-322.

Burman, R. D., J. L. Wright and M. E. Jensen. 1975. Changes in climate and estimated evapotranspiration across a large irrigated area in Idaho. *Transactions of the ASAE* 18(6):1089-1093.

Butler, G. D., Jr. and D. R. Scott. 1976. Two models for development of the corn earworm on sweet corn in Idaho. *Environmental Entomology* 5(1):68-72.

Croft, B. A., J. L. Howes and S. M. Welch. 1976. A computer based, extension pest management delivery system. *Environmental Entomology* 5(1):20-34.

Dale, R. F. 1981. Existing meteorological networks. In *Computer Techniques and Meteorological Data Applied to Problems of Agriculture and Forestry: A Workshop*, 56-63, ed. A. Weiss. Boston, MA: American Meteorological Society.

Eckersten, H. 1986. Simulated willow growth and transpiration: The effect of high and low resolution weather data. *Agriculture and Forest Meteorology* 38:289-306.

Fargo, W. S. and E. L. Bonjour. 1988. Developmental rate of the squash bug, *Anasa tritis* (Heteroptera: Coreidae), at constant temperature. *Environmental Entomology* 17(6):929-929.

Fritchen, L. J. and L. W. Gay. 1979. *Environmental Instrumentation*. New York: Springer-Verlag.

Fujioka, F. M. and M. A. Fosberg. 1981. Design of a weather station network for natural resource management. In *Computer Techniques and Meteorological Data Applied to Problems of Agriculture and Forestry: A Workshop*, 28-40, ed. A. Weiss, Boston, MA: American Meteorological Society.

Garcia-Salazar, C., H. Podoler and M. E. Whalon. 1988. Effects of temperature on diapause induction in the codling moth, *Cydia pomonella* (L.) (Lepidoptera: Olethreutidae). *Environmental Entomology* 17(4):626-628.

Gutierrez, A. P., D. W. DeMichele, Y. Wang, G. L. Curry, R. Skeith and L. G. Brown. 1980. The systems approach to research and decision making for cotton pest control. In *New Technology of Pest Control*, 155-186, ed. C. B. Huffaker. New York: John Wiley & Sons.

Harcourt, D. G., Y. M. Yee and J. C. Guppy. 1983. Two models for predicting the seasonal occurrence of *Agromyze frontella* (Diptera: Agromyzidae) in eastern Ontario. *Environmental Entomology* 12(5):1455-1458.

Hartstack, A. W., Jr., J. P. Hollingsworth, R. L. Ridgeway and J. R. Coppege. 1973. A population dynamics study of the bollworm and the tobacco budworm with light traps. *Environmental Entomology* 2(2):244-252.

Hatfield, J. L. and I. J. Thomason, ed. 1982. *Biometeorology in Integrated Pest Management*. New York: Academic Press.

Haynes, D. L., R. K. Brandenburg and P. D. Fisher. 1973. Environmental monitoring networks for pest management systems. *Environmental Entomology* 2(5):889-899.

Henneberry, T. J. and T. E. Clayton. 1983. Pink bollworm (*Lepidoptera: Gelechiidae*): Effects of soil moisture on behavior of diapausing larvae and adult emergence from bolls. *Environmental Entomology* 12(5):1490-1495.

Higley, L. G., L. P. Pedico and K. R. Ostlie. 1986. DEGDAY: A program for calculating degree-days and assumptions behind the degree day approach. *FORUM: Environmental Entomology* 15(5):999-1016.

Hurst, G. W. 1975. Meteorology of the Colorado Potato Beetle. WMO Tech Note No. 137. WMO 391. Geneva, Switzerland.

Jackson, R. D. 1982. Canopy temperature and crop water stress. In *Advances in Irrigation*, Vol. I, ed. D. Hillel, 43-85. New York: Academic Press.

Johnson, C. G. 1969. *Migration and Dispersal of Insects by Flight*. London: Methuen & Co. Ltd. Distrib. in USA by Barnes and Noble, New York.

Lindsey, A. A. and J. E. Newman. 1956. Use of official weather data in spring-time temperature analysis of an Indiana phenological record. *Ecology* 37:812-823.

List, R. J. 1949. *Smithsonian Meteorological Tables*. Smithsonian Misc. Coll., 6th Rev. Ed., Vol. 114. Washington, DC.

Logan, J. A., D. J. Wollkind, S. C. Hoyt and L. K. Tanigoshi. 1976. An analytic model for description of temperature dependent rate phenomena in arthropods. *Environmental Entomology* 5(6):1137-1140.

McCann, I. R. 1985. A soil temperature model for predicting the over-wintering emergence of *Heliothis zea* and *Heliothis virescens* (Lepidoptera: Noctuidae). Ph.D. diss., Texas A&M University, College Station.

Merriam, C. H. 1894. Laws of temperature continuity of the geographical distribution of terrestrial animals and plants. *National Geographic* VI:229-238.

Mitchell, J. M., Jr. 1958. Effect of changing observation time on mean temperature. *Bulletin of the American Meteorological Society* 39:83-89.

Omar, M. H. 1980. Meteorological factors affecting the epidemiology of the cotton leaf worm and the pink bollworm. WMO Tech Note No. 162. WMO 532. Geneva, Switzerland.

Parton, W. J. and J. A. Logan. 1981. A model for diurnal variation in soil and air temperature. *Agriculture Meteorology* 23:205-216.

Quaintance, A. L. and C. T. Brues. 1905. The cotton bollworm. Bureau of Entomology Bulletin No. 50, Washington, DC.

Rainey, R. C. 1974. Biometeorology and insect flight: Some aspects of energy exchange. *Annals Rev. of Entomology* 19:407-439.

Reichenbach, N. G. and G. R. Stairs. 1984. Response of the western spruce budworm, *Ahoristaneura occidentalis* (Lepidoptera: Tortricidae), to temperature: The stochastic nature of developmental rates and diapause termination. *Environmental Entomology* 13(6):1549-1556.

Reicosky, D. C., L. J. Winkelman and D. G. Baker. 1989. Accuracy of hourly air temperatures calculcated from daily minima and maxima. *Ag. and Forest Meteor.* 46:193-209.

Riley, C. V. 1880. The cotton worm. U.S. Ento. Commission Bulletin No. 3. Washington, DC: USDA.

Ring, D. R. and M. K. Harris. 1983. Predicting pecan nut casebearer (*Lepidoptera: Pyralidae*) activity at College Station, TX. *Environmental Entomology* 12(2):482-486.

Ring, D. R., V. R. Calcote and M. K. Harris. 1983. Verification and generalization of a degree-day model predicting pecan nut casebearer (*Lepidoptera: Pyralidae*) activity. *Environmental Entomology* 12(2):487-489.

Rodriguez-Iturbe, I. and J. M. Mejia. 1974. The design of rainfall networks in time and space. *Water Resources Research* 10(4):713-728.

Rosenberg, N. J., B. L. Blad and S. B. Verma. 1983. *Microclimate: The Biological Environment*, 2nd Ed. New York: John Wiley & Sons.

Rosenberg, N. J. and S. B. Verma. 1978. Extreme evapotranspiration by irrigated alfalfa: A consequence of the 1976 midwestern drought. *Journal of Applied Meteorology* 17:934-941.

Rumbaugh, W. F. 1934. The effect of time of observation on mean temperature. *Monthly Weather Review* 62:375-376.

Russell, J. S. 1984. Mathematical representation of related periodic variables using Fourier series and projections of vector functions: Application to the biological sciences. *Agriculture and Forest Meteorology* 33:215-224.

Sanderson, E. D. 1910. The relation of temperature to the growth of insects. *Journal of Econ. Entomology* 3(2):113-140.

Sevacherian, V., V. M. Stern and A. J. Mueller. 1977. Heat accumulation for timing Lygus control measures in a safflower-cotton complex. *Journal of Econ. Entomology* 70:399-402.

Stinner, T. E., R. L. Rabb and J. R. Bradley. 1973. Population dynamics of *Heliothis zea* (Boddie) and *H. virescens* (F.) in North Carolina: A simulation model. *Environmental Entomology* 3(1):163-168.

Snyder, R. L. 1985. Hand calculating degree days. *Agriculture and Forest Meteorology* 35:353-358.

Taylor, F. 1981. Ecology and evolution of physiological time in insects. *American Naturalist* 117(1):1-23.

Tauber, M. J., C. A. Tauber and S. Masaki. 1986. *Seasonal Adaptation of Insects*. New York: Oxford University Press.

Throne, J. E. and C. J. Eckenrode. 1986. Development rates for the seed maggots *Delia platura* and *d. florilega* (Diptera: Antonmyiidae). *Environmental Entomology* 15(5):1022-1027.

Uvarov, B. P. 1931. Insects and climate. *Transactions of the Entomological Society of London* 79:1-247.

Wagner, T. L., H. Wu, P. J. H. Sharpe, R. M. Schoolfield and R. N. Coulsen. 1984. Modeling insect development rates: A literature review and application of a biophysical model. *Annals of the Entomological Society of America* 77:208-225.

Wagner, T. L., R. O. Flamm, H. Wu, W. S. Fargo and R. N. Coulsen. 1987. Temperature-dependent model of life cycle development of *Ips Calligraphus* (Coleoptera: Scolytidae). *Environmental Entomology* 16(2):497-502.

Wang, J. Y. and R. K. Ray. 1984. *Agriculture and its Environment: Prediction and Control.* Dubuque, IA: Kendall Hunt Publishing.

Wann, M., D. Yen and J. H. Gold. 1985. Evaluation and calibration of three models for daily cycle of air temperature. *Agriculture and Forest Meteorology* 34:121-128.

Wellington, W. G. and R. M. Trimble. 1984. Weather. In *Ecological Entomology,* 399-425, ed. C. B. Huffaker and R. L. Rabb. New York: John Wiley & Sons.

Yao, A. Y. M. 1981. Agricultural climatology, Ch. 2. In General Climatology, Vol. 3, of *World Survey in Climatology*, ed. H. E. Landsberg, 189-298. Amsterdam: Elsevier.

7

Near-surface Soil Temperature Model for Biophysical Development Models

Ian R. McCann, Marshall J. McFarland, and Jack A. Witz

Introduction

A basic assumption made in biophysical models is that the rate of development of an organism is a function of the physical environment in which the organism is located, in particular the thermal environment. For organisms which spend all or part of their lives in the soil, a model for soil temperature as it varies with time is an important component of a biophysical development model. Such organisms, insects for example, are often located near the soil surface, where temperature is determined by complex and dynamic fluxes of matter and energy. A biophysical development model for such organisms can be no more accurate than the soil temperature component of the model, but an accurate model of temperatures near the soil surface requires data that is not generally available in type and quantity. Models of soil temperature have been developed for a variety of purposes and with varying degrees of complexity and data requirements. They range in approach from the more physical and deterministic models (Hanks et al., 1971; Schroeder et al., 1978) to the more empirical and statistical models (Cruse et al., 1980; Gupta et al., 1981). In order to be useful in a biophysical model with practical applications, a model should require only data that is widely available, and it should make simplifying assumptions where necessary to allow for data limitations and computing capabilities.

A common assumption in simple soil temperature models is that the effect of soil water content can be ignored. This assumption may be reasonable for models in which the temperature regime near the surface is not of major concern, but may not be valid where near surface temperatures are important because soil water and associated energy fluxes at or near the surface are generally greater than elsewhere in the profile.

This chapter presents a near-surface soil temperature model for bare soils which incorporates estimates of soil water content and its effect on soil temperature. Daily meteorological data is required for solar radiation, humidity, air temperature, wind speed and rainfall or irrigation, and hourly estimates of soil temperature over time are generated for a specified depth. These estimates may be integrated with biophysical models for which soil temperature near the surface is required, such as in overwintering insect development models (McCann, 1985).

Model Description

Heat energy will flow from one location to another if the two locations are at different temperatures and if there is a thermally conducting medium separating them. The magnitude of the heat flux depends upon the temperature gradient and the thermal conductivity. The temperature that results from the input to or removal of heat energy from a given volume of material depends upon the magnitude of the energy change and the thermal capacity of the material. In a soil, the temperature gradient varies with time as the surface heats or cools, and both the thermal conductivity and thermal capacity vary as the water content and temperature of the soil varies.

An equation describing soil temperature as it varies with depth and time is:

$$C_v(\theta)\frac{\partial T}{\partial t} = \frac{\partial}{\partial z}\left[K_T(\theta,T)\frac{\partial T}{\partial z}\right] \qquad (7.1)$$

where

T = temperature (°C)
t = time (s)
K_T = thermal conductivity (W m^{-1} °C^{-1})
C_v = thermal capacity (J m^{-3} °C^{-1})
z = distance downwards from surface (m)
θ = volumetric water content (m^3 m^{-3})

Equation 7.1 cannot be solved analytically unless assumptions are made about thermal conductivity, thermal capacity and the boundary conditions.

deVries (1963) developed equations to estimate the bulk thermal conductivity and thermal capacity of soils from the corresponding properties and respective volume fractions of the solid, liquid and gaseous phases. van Bavel and Lascano (1979) developed a physically based numerical model named CONSERVB which used the methods of deVries (1963) to estimate thermal conductivity and thermal capacity and which solved equation 7.1 for bare soils. In this model the soil profile is divided into discrete layers, each of which may have different properties. Temperature and volumetric water content for each of the layers is estimated for each time step. The upper boundary condition, the soil surface temperature over time, is estimated using the surface energy balance, with the latent energy component (evaporation) estimated using an aerodynamic resistance equation. Water movement within the soil profile is estimated from both the hydraulic conductivity and the pressure potential relationships to volumetric water content. CONSERVB has been experimentally verified in the College Station, Texas, area (Lascano and van Bavel, 1983) and

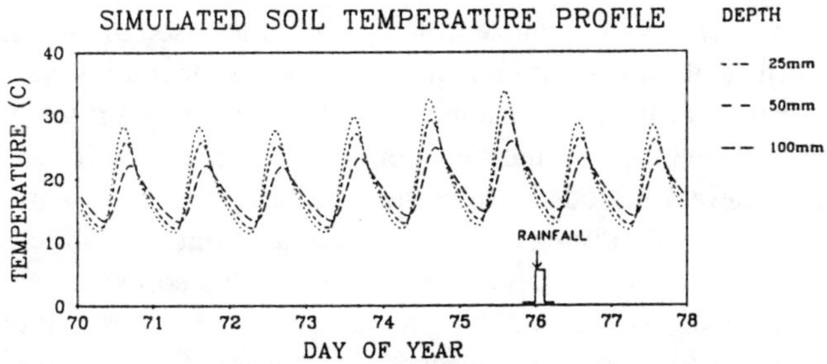

Figure 7.1–Simulated soil temperature profile before and after a rainfall event using CONSERVB with simulated input data.

by McCann (1985) using a version of CONSERVB modified to run on a minicomputer.

The modified version of CONSERVB was used to investigate the relative importance of the physical processes which determine soil temperature near the surface, and in particular the effect of soil water on these processes. The two principal physical mechanisms in which water content is involved in determining near surface soil temperatures are:

1. The conduction of heat to and from the surface and the storage of heat within the soil profile.
2. The surface energy balance and resulting surface temperature.

The first of these primarily relates to the bulk of the soil, with soil water affecting the bulk thermal and hydraulic properties. The second primarily relates to the soil surface, with soil water availability affecting the latent heat component of the surface energy balance and, consequently, the surface temperature. The effect of soil water content on each of these processes as they relate to soil temperature was investigated with a series of simulations. Simulated input data was used to eliminate the random variability inherent in real data while at the same time preserving typical diurnal variations and trends over time. A series of simulated data sets were generated using appropriate sine functions or constants for hourly air temperature, solar radiation, dew point temperature, and wind speed. Figure 7.1 shows simulated soil temperatures in response to a diurnal meteorological regime that is the same for each day of the simulation, with the exception of one rainfall event to rewet the soil. It can be seen that, as the soil dries, the temperature maxima increase. Following the rainfall, the temperature maxima decrease to approximately the level at the start of the period.

The effect of soil water content on temperature due to effects on bulk soil thermal properties was investigated in a series of simulations in which soil water content and distribution in the profile were held constant over time at typically dry and typically wet values. Surface temperature at any given time was essentially the same for both cases because, even in the dry soil simulations, the surface water content was above the level at which evaporation is inhibited and the surface energy balance altered. The results showed that there was little difference in soil temperatures between the wet and dry soil at any given time for any of the meteorological regimes used in the simulations.

The effect of soil water content on surface temperature was investigated with another series of simulations in which the soil at the surface was permitted to dry below the value at which evaporation becomes significantly inhibited. This causes the estimated daytime surface temperatures to increase under otherwise identical conditions, and this increase tends to directly effect soil temperatures near the surface. Soils with higher initial soil water contents took a longer time to reach the point at which surface temperatures rise than the dryer soils did, but once this point was reached, there was little difference between the surface temperatures. The effect of a rainfall event was to immediately suppress the rise until the surface was sufficiently dry again.

Figures 7.2a and 7.2b show estimated air temperature, surface temperature and latent energy flux for a drying soil. It can be seen that, as the soil dries and the latent energy flux due to evaporation decreases, daytime surface temperatures increase above air temperatures. The daily surface temperature maximum is considerably higher than the air temperature maximum when the soil surface is dry, while when the surface is wet the maxima are approximately the same.

If the effect of soil water content on the bulk thermal properties as they affect near surface temperatures is relatively small, then the soil may be thought of as a thick homogeneous "slab", with thermal capacity and thermal conductivity assumed constant with depth and time. Furthermore, if a diurnal sinusoidally varying surface temperature is assumed, the solution to equation 7.1 is given by:

$$T(z,t) = T_{avg} + T_{amp}(0)e^{-B}\,\mathrm{SIN}(\omega t - B) \tag{7.2}$$

where

z	=	depth below surface (m)
t	=	time (s)
$T_{amp}(0)$	=	amplitude of the temperature sine function at the surface (°C),
	=	$[\,T_{max}(0) - T_{min}(0)\,]/2$
$T_{max}(0)$	=	daily maximum temperature at surface (°C)
$T_{min}(0)$	=	daily minimum temperature at surface (°C)
T_{avg}	=	average soil temperature (°C)
		$[\,T_{max}(0) + T_{min}(0)\,]/2$
ω	=	angular frequency of the diurnal sine wave (s^{-1})

and

$$B = z\left(\frac{\omega}{2D_T}\right)^{0.5} \tag{7.3}$$

where D_T is the thermal diffusivity ($m^2\ s^{-1}$), defined as $[K_T/C_v]$.

The average soil temperature in this solution is constant with depth, while the amplitude of the surface temperature sine wave decreases exponentially with depth, as indicated by the exponential term in equation 7.2. The phase shift

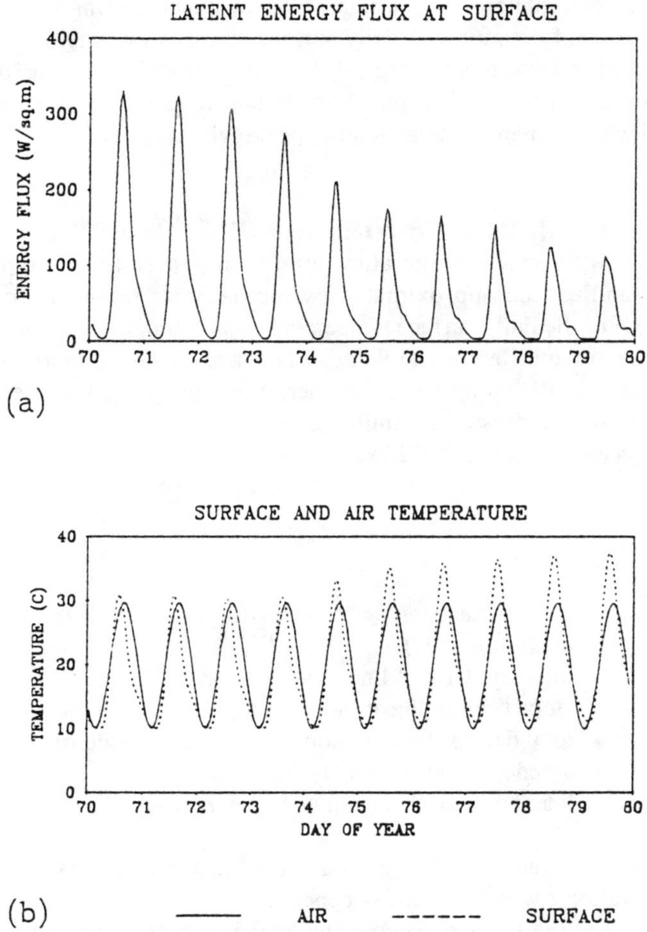

Figure 7.2–Latent energy flux for a drying soil estimated using (a) CONSERVB with simulated input data, and (b) associated surface and air temperatures.

(lag) increases with depth as indicated by the sine term, but for biophysical development models this is often unimportant.

These results were used to develop a simple model of soil temperature near the surface that includes the influence of soil water content through its effect on surface temperatures, but which assumes that the effect on bulk thermal properties may be neglected. The principal steps in the model are:

1. Estimation of the daily surface temperature maxima and minima.
2. Simulating a daily surface temperature "wave" by interpolation between the daily surface temperature maxima and minima.
3. Propagation of this wave downwards into the soil to the desired depth.

A series of numerical experiments was conducted using CONSERVB to derive equations for estimating daily surface temperature maxima and minima from widely available meteorological data, and for estimating the decrease of temperature amplitude with depth. Simulations were run for a variety of soil types, soil water contents, and meteorological regimes.

Surface Temperature Maxima and Minima

The daily surface temperature minimum generally occurs during the nighttime and can be approximated by the daily air temperature minimum (fig. 7.2b). For the daily surface temperature maximum, a good correlation was found between the difference in daily surface and air temperature maxima and the difference in incoming net solar energy and outgoing latent energy from evaporation (both expressed in similar units). The resulting equation, which had a correlation coefficient of 0.972, was:

$$\Delta T_{max} = a + b\left[R_{sol}(1-\alpha) - H_{ev}\right] \tag{7.4}$$

where

ΔT_{max}	=	daily surface temperature maximum minus air temperature maximum (° C)
a	=	intercept term in linear model, –2.96 (° C)
b	=	slope term in linear model, 3.5 (° C/mm water equivalent)
R_{sol}	=	total daily solar radiation (mm water equivalent)
α	=	albedo of soil surface
H_{ev}	=	latent energy flux (mm water equivalent)

The use of equation 7.4 requires an estimate of the daily latent energy flux (soil evaporation), which in turn is dependent upon both the weather and soil water content at the surface. Ritchie and Burnett (1971) described a model for soil evaporation based upon the assumption that evaporation from a bare soil takes place in two stages. During the first stage the soil is sufficiently wet to permit evaporation to occur at the potential rate, as determined by atmospheric demand. The daily surface temperature maximum is approximately equal to the air temperature maximum under these conditions (figs. 7.2a and 7.2b). As the soil dries, evaporation will proceed at the potential rate until the ability of the soil to conduct water to the surface becomes the limiting factor. At this point evaporation enters the second stage and is reduced below the potential rate. It is during the second stage that the daily surface temperature maximum will tend to increase above the air temperature maximum. In the simple model, soil evaporation is assumed to be equal to potential evaporation during the first stage and a fraction of potential evaporation during the second stage, with the fraction being determined using a function of the number of days since the second stage commenced (Ritchie and Jordan, 1972; Bordovsky, 1978). Daily potential evaporation was estimated using the Penman method (Doorenbos and Pruitt, 1977), and daily soil water content was estimated using a layer model

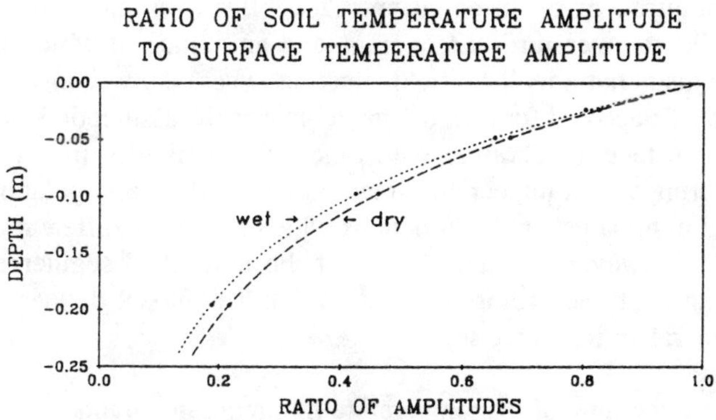

Figure 7.3–Ratio of soil temperature amplitude to estimated surface temperature amplitude, as a function of depth, for a wet and dry soil.

developed by Bordovsky (1978). The second stage is assumed to commence when the daily soil water content in the layer closest to the surface falls below a specified fraction of field capacity. Below this average value, the soil is not able to conduct water to the surface sufficiently rapidly to meet atmospheric demand. A value of 75% of field capacity for this fraction gave the best results for the Norwood silt loam used to verify CONSERVB. The daily surface temperature maxima and minima can be used to calculate the surface temperature amplitude. The daily soil temperature amplitude at a given depth is then estimated from the surface temperature amplitude.

Soil Temperature Amplitude

Figure 7.3 shows the average of the ratios of daily soil temperature amplitude to daily surface temperature amplitude, at 25, 50, 100, and 200 mm depths, for a 20-day simulation, for both a wet and a dry soil in which hourly surface temperatures were the same. The scatter of points around the average at each depth was small, and it can be seen that an exponential curve fits the four averages for each of the wet and the dry soils very well. This is consistent with equation 7.2 in which the temperature amplitude decreases exponentially downwards from the surface. Furthermore, the two curves are close together, which is consistent with the conclusion that the effect of soil water content on the bulk thermal properties of a soil may be neglected in a simple model. The exponential function corresponding to the mean of the two curves is:

$$\frac{T_{amp}(z)}{T_{amp}(0)} = e^{-8.23z} \tag{7.5}$$

where $T_{amp}(z)$ is the temperature amplitude at depth z (° C) and $T_{amp}(0)$ is the temperature amplitude at surface (°C).

The solution given in equation 7.2 applies to a sinusoidally varying diurnal surface temperature, the average and amplitude of which is constant over time. This is not a realistic field situation, where daily changes in weather are the norm. To account for this, the model makes the assumption that the daily variation of surface temperature with time may be divided into a number of segments during which temperature may be modeled by interpolating between maxima and minima using sine functions which may have different amplitudes and periods. However, the amplitudes of the individual segments will still decrease with depth according to equation 7.4. The model divides the diurnal surface temperature into three segments:

1. From the time of the temperature minimum in the morning to the time of the temperature maximum in the afternoon.
2. From the time of the temperature maximum in the afternoon to sunset.
3. From sunset to the time of the temperature minimum the following morning.

The sine functions for each segment were chosen so that at the times of temperature maxima and minima the derivative with respect to time is zero. This provides smooth transitions between segments and between adjacent days, with the exception of the transition at sunset. The daily surface temperature "wave" is then propagated down into the soil as given by equation 7.2, with the exponential decay component estimated by equation 7.4 to give a diurnal soil temperature "wave" for any given depth (McCann, 1985).

The average surface and soil temperatures on any day is defined as the average during segment 1. Because the daily surface and soil temperature averages are unlikely to be equal and constant over time, as they are in equation 7.2, the soil temperature "wave" is modified to allow for thermal inertia during periods of warming and cooling, when the daily surface temperature averages are different. The estimated minimum soil temperature in the morning is adjusted using a lag function of the difference between the soil and surface temperature averages during the previous day (eq. 7.6), with segment 3 of the diurnal "wave" correspondingly adjusted (McCann, 1985).

$$T_{min,\,z,\,i} = T_{avg,\,z,\,i-1} - \left[e^{-8.23\,z}\right]\left[T_{avg,\,s,\,i-1} - T_{min,\,s,\,i}\right] + L_T\left[T_{avg,\,s,\,i-1} - T_{avg,\,z,\,i-1}\right] \tag{7.6}$$

where

$T_{min,\,z,\,i}$ = minimum temperature at depth z on day i (° C)
$T_{avg,\,z,\,i-1}$ = average temperature at depth z on day i – 1 (° C)
$T_{amp,z}$ = temperature amplitude at depth z (° C)
$T_{amp,s}$ = temperature amplitude at surface (° C)
$T_{avg,\,s,\,i-1}$ = average temperature at surface on day i – 1 (° C)
$T_{min,\,s,\,i}$ = minimum temperature at surface on day i (° C)
L_T = lag constant for temperature difference

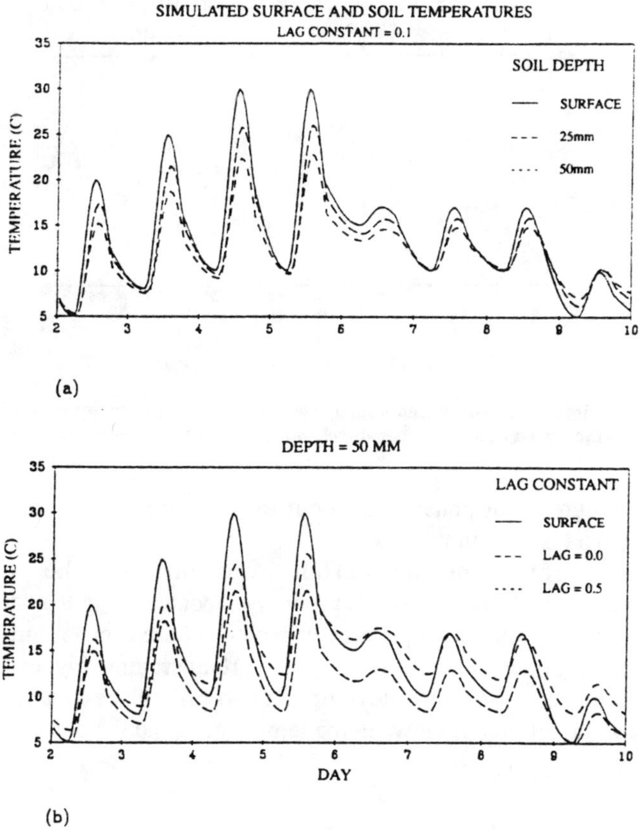

Figure 7.4–Response of soil temperature at 25 and 50 mm to a simulated surface temperature regime, using a lag constant of 0.1. (a), and at 50 mm using lag constants of 0.0 and 0.5. (b).

Figure 7.4a shows estimated soil temperature at 25 mm and 50 mm, using a lag constant of 0.1, in response to simulated diurnal surface temperatures which exhibit periods of increasing and decreasing maxima and minima. Figure 7.4b shows the estimated soil temperature at 50 mm in response to the same surface temperatures, but with lag constants of 0 and 0.5. The higher the lag constant the more influence the previous day's surface temperature temperature maximum has. A value of 0.1 for the lag constant was found to give the best results.

Results and Discussion

Figure 7.5 shows a typical comparison between simulated and measured hourly soil temperature at 25 mm for a 10-day period in March 1984. There is

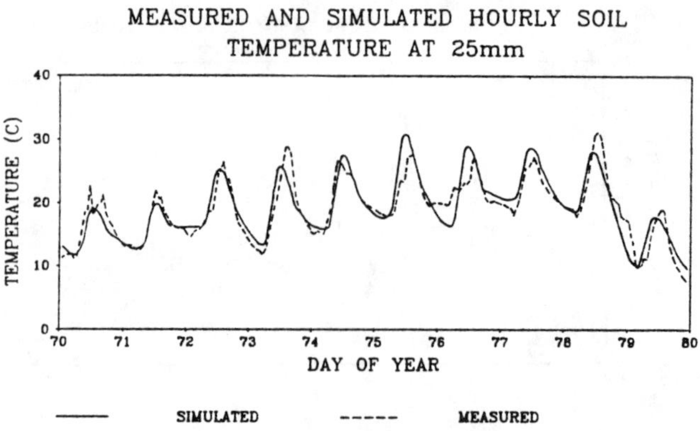

Figure 7.5–Simulated soil temperature at 25mm compared with measured hourly temperatures, for a 10-day period in early spring.

reasonable agreement considering both the proximity to the surface and that only daily data is used in the model.

The effect of the thermal environment upon the physiological development rate of organisms is commonly modeled using the "degree-day" or "heat unit" accumulation approach, in which a base temperature is identified below which no development takes place. Temperatures higher than the base over a period of time cause development to take place at a rate which is a function of the difference between the temperature and the base.

$$HU_i = \frac{(T_i - T_B)^P}{24}, \quad T_i \geq T_B$$

$$HU_i = 0, \qquad T_i < T_B \tag{7.7}$$

where

HU_i = heat units available during hour i (degree-days)
T_i = temperature during hour i (° C)
T_B = base temperature (° C)
P = exponent to account for non linearity of a biological response

In equation 7.7, heat units are calculated hourly and converted to units of "degree-days." Cumulative physiological development during a period of time requires summing heat units for the period (eq. 7.8).

$$D_{ab} = \sum_{i=a}^{i=b} HU_i \tag{7.8}$$

where

D_{ab} = biological development between times a and b
HU_i = heat units available during period i

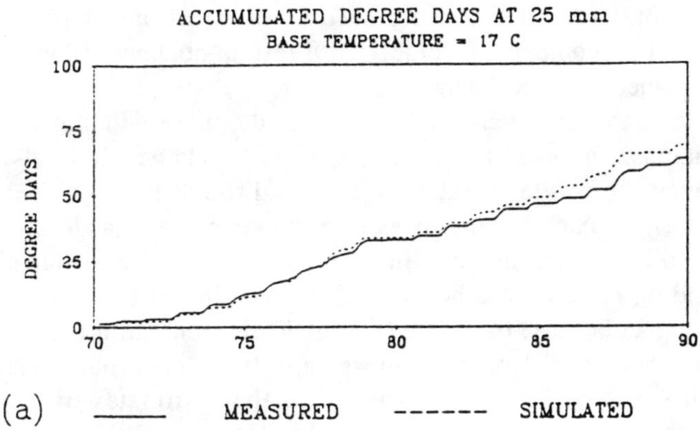

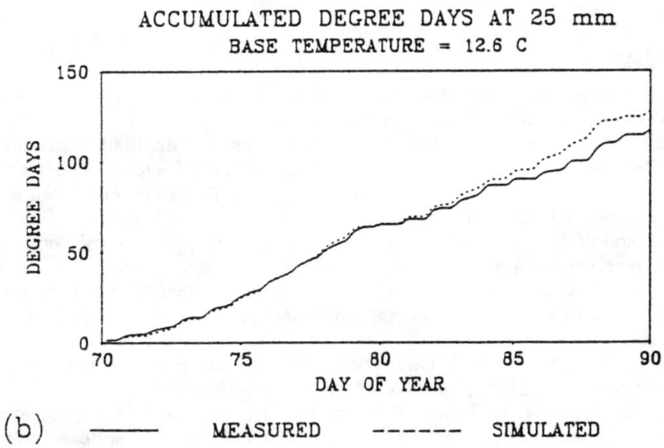

Figure 7.6–Observed and estimated hourly heat unit accumulation at 25mm, for base temperatures of 17°C (a) and 12.6°C (b), for a 20-day period in early spring.

The measured rate of heat unit accumulation at 25 mm is compared with the simulated rate in figure 7.6, for base temperatures of 17.0 and 12.6° C. These are representative of pupal depth and base temperatures for diapause development and pupation, respectively, for overwintering *Heliothis zea* (Cotton Bollworm) (Lopez and Hartstack, 1983; Lopez and Hartstack, 1985). For comparison purposes a linear function is used to calculate heat units as a function of temperature (P = 1.0 in eq. 7.7). It can be seen that the agreement during the period of measurement is reasonably close, with the simulations tending to overestimate accumulated heat units. With increasing depth, however, the simulations tend to underestimate the observed total slightly. The magnitude of the difference after 20 days however was less than 10% of the measured total for all the combinations of depth and base temperature considered. This compares favorably with results obtained using CONSERVB with hourly meteorological data.

Of the habitats in which organisms reside, those with a bare soil surface are the simplest to model in physical terms. The relative effect of a vegetative canopy upon the soil temperature regime is difficult to model because of the complexities the canopy introduces into the energy exchange processes. The daytime soil temperature maxima are likely to be less than those for a corresponding bare surface because of the shading effect, while the minima might be higher because of reduced energy losses at night. It is likely, however, that development will proceed more rapidly under a bare surface than a vegetated surface, with the consequence that estimates of physiological development produced by this model, for any local population of organisms, are likely to represent the individuals located in the most favorable micro-environments.

References

Bordovsky, J. P. 1978. Predicting farm machinery operation time with a soil moisture model. M.S. thesis. Texas A&M University, College Station.

Cruse, R. M., D. R. Linden, J. K. Radke, W. E. Larsen and K. Larntz. 1980. A model to predict tillage effects on soil temperature. *Journal of the Soil Science Society of America* 44:378-383.

deVries, D. A. 1963. Thermal properties of soils. In *Physics of the Plant Environment*, ed. van Wijk. Amsterdam, Netherlands: North Holland.

Doorenbos, J. and W. O. Pruitt. 1977. Crop water requirements. Irrigation and drainage paper 24. Rome, Italy: Food and Agriculture Organization of the United Nations.

Gupta, S. C., J. K. Radke and W. E. Larsen. 1981. Predicting temperatures of bare and residue covered soil with and without a corn crop. *Journal of the Soil Science Society of America* 45:405-412.

Hanks, R. J., D. D. Austin and W. T. Ondrechen. 1971. Soil temperature estimation by a numerical method. *Journal of the Soil Science Society of America* 35:665-667.

Lascano, R. J. and C. H. M. van Bavel. 1983. Experimental verification of a model to predict soil moisture and temperature profiles. *Journal Soil Science Society of America* 47:441-448.

Lopez, J. D. and A. W. Hartstack. 1983. Diapause development of *Heliothis* spp. *Journal of the Georgia Entomological Society* 18(1):104-111.

Lopez, J. D. and A. W. Hartstack. 1985. Comparison of diapause development in *Heliothis zea* and *H. virescens* (Lepidoptera: Noctuidae). *Annals of the Entomological Society of America* 78:415-422.

McCann, I. R. 1985. A soil temperature model for predicting the overwintering emergence of *Heliothis zea* and *Heliothis virescens* (Lepidoptera: Noctuidae). Ph.D. diss. Texas A&M University, College Station.

Ritchie, J. T. and E. Burnett. 1971. Dryland evaporative flux in a subhumid climate: II. Plant influences. *Agronomy Journal* 63:56-62.

Ritchie, J. T. and W. R. Jordan. 1972. Dryland evaporative flux in a subhumid climate: IV. Relation to plant water status. *Agronomy Journal* 64:173-176.

Schroeder, C., D. W. Michele, V. W. Pooch and W. R. Teague. 1978. Modeling heat and moisture flow in soils. *Simulation* (November):173-179.

van Bavel, C. H. M. and R. Lascano. 1979. CONSERVB. A numerical method to compute soil water content and temperature profiles under a bare surface. Texas A&M University, Remote Sensing Center.

PART II

INSECT POPULATION DYNAMICS

8

Representation of Development in Insect Population Models

Glen H. Smerage

Introduction

Models are important analytical tools for predicting, evaluating, and understanding the dynamics of populations of pest and beneficial organisms in natural and man-made ecosystems under a variety of environmental conditions and management actions. Need for better capability to predict and understand insect population dynamics has motivated a quest for better models. Early population models by Verhulst (1838), Pearl and Reed (1920), Lotka (1924), Volterra (1926), and Nicholson and Bailey (1935), which are best described as dynamic curve fitting, no longer suffice.

The thrust over the past 40 years has increasingly emphasized more detailed, mechanistic models of population systems developed by analyzing relevant biological and ecological processes comprising such systems. Leslie (1945, 1948) first pursued this objective, and others have carried it forward by various approaches (Oldfield, 1966; Fredrickson et al., 1967; Sinko and Streifer, 1967; Weiss, 1968; Langhaar, 1972; Caswell, 1972; Auslander et al., 1974; Tummala et al., 1975; Gutierrez et al., 1980; Shoemaker, 1980; Curry and Feldman, 1987).

A contemporary view of insect populations is illustrated by the simple conceptual model in figure 8.1, which depicts a system of two species interacting through predation. The principal processes in the system are development, reproduction, physiological and predatory mortalities, and migration. Egg, larval, pupal, and adult stages and reproduction are shown for each species; in each case, reproduction produces a flow of new eggs in proportion to the adult population. After oviposition, individuals move through the life cycle, remaining in each stage for a random interval commensurate with development in that stage and the prevailing temperature regime. Physiological mortality removes adults and some immatures of each species. Mortality of prey larvae is primarily by predation at a rate depending on adult predator and larval prey population densities. Adults of both species may migrate across the system boundary defining the region occupied by the two populations.

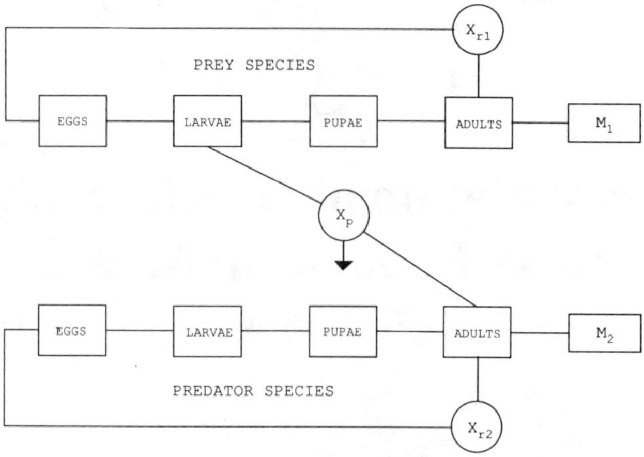

Figure 8.1–Block diagram of two insect species interacting through predation, X_p. Processes X_{r1} and X_{r2} represent reproduction; M_1 and M_2 represent migration.

Modeling is the task of describing the processes in and structure of such a conceptual model and subsequently formulating from those descriptions a mathematical model representing the real system. Definition must be given to each box and circle in a conceptual model (e.g., fig. 8.1). One approach is rooted in convolution techniques; another is based on identification and description of processes. In the latter, objects may be described in terms of four genera of elementary processes—storage, transport, transformation, and source—whose unique behavioral properties contribute to the behavioral properties of any system comprised of them (Smerage, 1985). The first significant realization is that a population is a type of process, namely storage. Reproduction may be represented by a transformation or controlled source process, depending on the organism. Mortality, including predation and parasitism, is a transformation, and migration involves transport mechanisms.

Insect development is a compound process, and representation of it in population models is the topic of this chapter. In particular, structure will be given to the boxes in figure 8.1 representing life stages by proceeding from the postulate that development time and mortality are the features of development in a life stage essential to population dynamics. Two classes of models, convolution and network, are proffered for representing individual and successive stages of life cycles. Convolution models have theoretical value, while network models are preferred in practice. Complexity and parameter values of models in either class are determined from the statistics of development time and survivorship in a stage represented. All variables are considered to be continuous functions of time.

Development

Development of an insect over its life cycle is a complex biochemical process characterized by four distinct stages: egg, larval, pupal, adult. Sharpe, DeMichele, Curry, and Feldman are among recent contributors to current understanding of the dynamic, biochemical basis of insect development (Sharpe and DeMichele, 1977; Sharpe et al., 1977; Curry et al., 1978). The approach here treats development at a higher level as one of several processes in a complete population system.

Of several biological attributes of development in a life stage, development (residence or transit) time, its reciprocal, development rate, and mortality or survivorship are paramount to representing the stage in a population model. This interval between entering and leaving a stage is affected by temperature and, to some degree, humidity. Modern biological studies of insects often include determination of development times in life stages and the dependence of them on temperature and humidity. Only effects of temperature are considered here.

Development time, t, of an insect in a life stage is a random variable with statistics determined by the following general experiment. A large cohort of a specific species corresponding to the beginning of the specified life stage is established in an isothermal environment. The time interval for each survivor to emerge from the stage is recorded as its development time (also known as residence or transit time) in that stage. The relative frequency distribution, mean, and variance of survivor times for the cohort are calculated from the data as estimates of the true probability density function, mean, and variance. Repeating the experiment at several temperatures over a range (e.g., 10-35° C) establishes the effect of temperature on development. Analogous experiments are performed to establish the dependence on humidity, nutrition or another factor.

A hypothetical family of isothermal probability density functions for development time is presented in figure 8.2. Corresponding graphs of mean development time, t_d, and its reciprocal, development rate, r_d, as functions of temperature are typified by the solid line graphs in figure 8.3, and it generally is assumed that isothermal development occurs at uniform rate over the whole time interval. Both graphs exhibit threshold temperature T_0 (e.g., 10-12° C), below which development is arrested and above which normal development proceeds fairly linearly until temperature becomes too high. The dashed graphs in figure 8.3 correspond to linear dependence of development rate on temperature in which, for $T > T_0$:

$$r_d = \frac{1}{K}(T - T_0) \tag{8.1}$$

Additional information about the temperature dependence of development may be found in Sharpe and DeMichele (1977), Sharpe et al. (1977), and Curry and Feldman (1987).

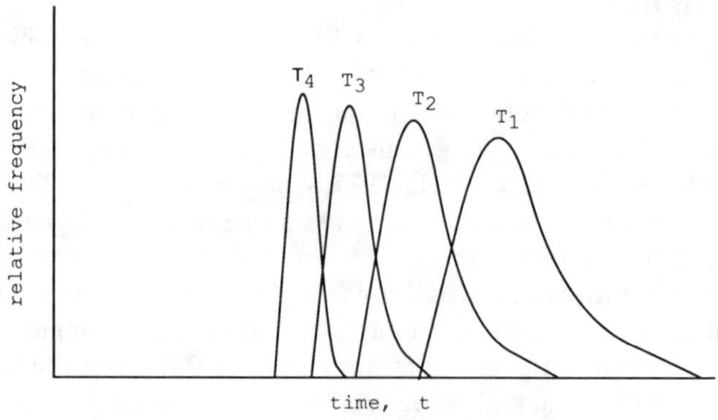

Figure 8.2–Hypothetical probability density functions of development time in an insect life stage for constant temperatures $T_4 > T_3 > T_2 > T_1$.

Information about physiological mortality in any immature stage is an adjunct to development time obtained from the experiment described above. Since time of death in an immature stage is difficult or impossible to establish, physiological mortality in a stage usually is accounted for by survivorship at emergence from the stage. At each temperature, the relative frequency of cohort survivors provides an estimate of the probability of stage survivorship, q, at the stated temperature; the corresponding probability of death, p, is given by $p = 1 - q$.

Treatment of the adult stage is somewhat different because all adults die, and, therefore, $p = 1$ and $q = 0$. Since departure from that stage corresponds to death, development time is better interpreted as time until death or adult residence time or longevity. The relative frequency distribution, mean, and variance of adult longevity are calculated from data obtained under stated conditions by experiments of the sort described above with adult cohorts.

The experiment outlined above may be interpreted as follows with the concepts of system analysis. Let the box in figure 8.4a depict a stage (e.g., egg, larval, pupal, adult) in an insect life cycle as a system with input, output, and state; it may similarly represent a sequence of stages. Input γ_i pertains to the flow of individuals into the stage, as from the preceding stage or, in the case of egg, from reproduction; output γ_o pertains to the flow of individuals from the stage to the next stage in the case of immatures or to the sink for the dead in the case of adults. A cohort of M individuals entering the stage is equivalent to exciting the system with an impulse (depicted by a vertical arrow at the input in fig. 8.4) of value M (Papoulis, 1980; Director and Rohrer, 1972; Cooper and McGillem, 1967; Lathi, 1965). In an isothermal environment, departure of survivors in the cohort from the stage at random times, t, corresponds to appearance at the system output of a delayed, finite-width pulse. With q denoting survivorship of the stage, the area of the output pulse

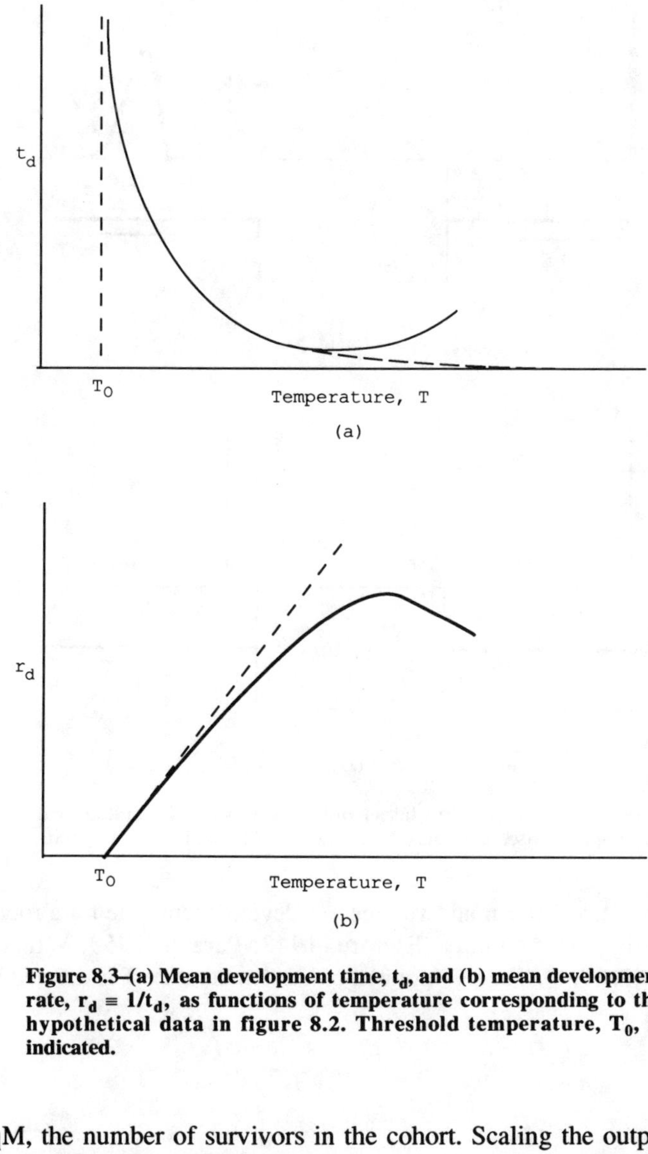

Figure 8.3–(a) Mean development time, t_d, and (b) mean development rate, $r_d \equiv 1/t_d$, as functions of temperature corresponding to the hypothetical data in figure 8.2. Threshold temperature, T_0, is indicated.

equals qM, the number of survivors in the cohort. Scaling the output pulse by 1/qM yields impulse response h of the stage (ibid.); h is equivalent to the isothermal probability density function of survivor development time (fig. 8.2 for specified temperature). The conclusions are: (1) an appropriate system representation of the temporal aspect of development in a life stage is dispersive delay; and (2) experimental data on development in a stage are directly applicable to formulation of a system model of that stage. In the sequel, h will synonymously denote the impulse response of a life stage and the probability density function of development time in that stage.

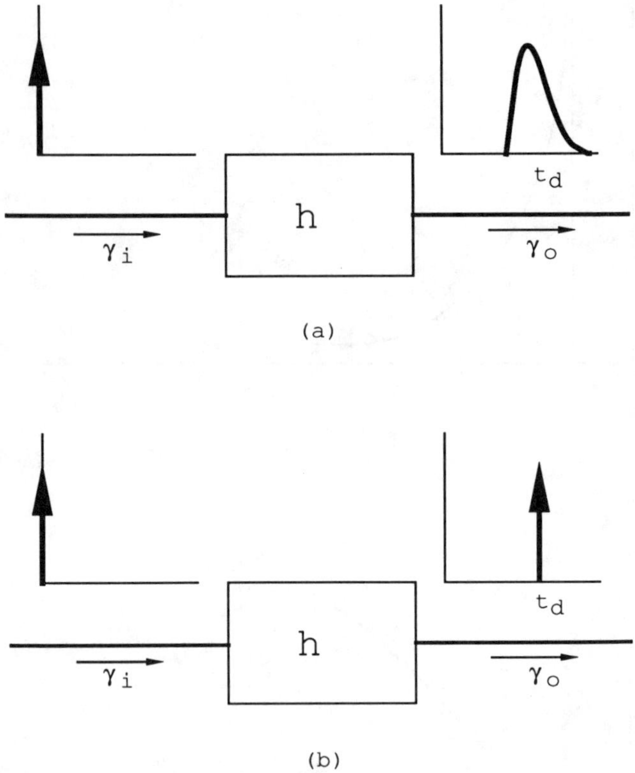

Figure 8.4–A stage of an insect life cycle viewed as a system with impulse response, h, that is (a) dispersive or (b) ideal delay of t_d units.

The isothermal mean and variance of development time t are found from h by the theory of moments (Elmore, 1948; Parzen, 1960; Wilks, 1967; Papoulis, 1962, 1984). The n*th* moment of h at temperature T is defined by:

$$m_n\,(T) = \int_0^{\infty} t^n\, h(t, T)\, d\,t \tag{8.2}$$

Note that m_0 equals the area under h.

$$m_0 = \int_0^{\infty} h(t)\ dt \tag{8.3}$$

It follows that the mean and mean square of, i.e., mean development time and mean square development time, t, denoted by $< t >$ and $< t^2 >$, respectively, are given by:

$$<t> \equiv t_d = \frac{m_1}{m_0} \tag{8.4}$$

and

$$<t^2> = \frac{m_2}{m_0} \tag{8.5}$$

Finally, the variance, σ_d^2, of, i.e., development time, t is given by:

$$\sigma_d^2 = \frac{\int_0^\infty (t - <t>)^2 h(t)\, dt}{m_0} \tag{8.6}$$

and coefficient of variance CV is given by (Wilks, 1967):

$$CV \equiv \frac{\sigma_d}{t_d} \tag{8.7}$$

Statistics of Development

Although the statistics of development are determined for each species by experiments described above, one naturally asks if probability density functions of development time have a functional form common to most species. The Erlang density function is used often, and Wagner et al. (1985) successfully fitted relative frequency distributions of development time with the Weibull function. Gamma, log-normal, and inverse normal functions also have been employed. Otherwise, there appear to be no suggestions in the literature for a standard density function from theoretical or empirical findings. Nevertheless, it is generally recognized that density functions of isothermal development time for various species are similar in shape and asymmetric about their mean times.

Because this discussion focuses on model formulation, only the ideal, Erlang, and Bessel density functions are considered. Some of the analyses involve Laplace transforms (Papoulis, 1980; Director and Rohrer, 1972; Cooper and McGillem, 1967; Lathi, 1965). Particularly important is the transform, denoted by H, of impulse response h. It also is known as the transfer function of the system since it equals the ratio of Laplace transforms of the system output γ_o and input γ_i.

Ideal Development. In ideal development, a cohort passes without dispersion (i.e., $\sigma_d = 0$) through a life stage in t_d units of time, and it follows that density function h_I of ideal development time t is a unit impulse delayed by t_d (ibid.), as shown in figure 8.4b.

$$h_I(t) = \delta(t - t_d) \tag{8.8}$$

where $\delta(t)$ denotes the unit impulse occurring at $t = 0$. The Laplace transform of h_I is:

$$H_I(s) = e^{-t_d s} \tag{8.9}$$

where s is the complex variable of Laplace Transforms (ibid.). It can be proven (Papoulis, 1980; Director and Rohrer, 1972; Cooper and McGillem, 1967; Lathi, 1965) that the temporal response of an ideal stage to general input γ_i is that input delayed by t_d. That is:

$$\gamma_o(t) = \gamma_i(t - t_d) \tag{8.10}$$

Erlang Description of Development. Development time of a life stage in fact exhibits dispersion as well as delay, as in figure 8.4a. The Erlang function (Manetsch, 1976; Gross and Harris, 1974; Cooper, 1972) with parameters m, the order of the function, and t_d, the delay, often is used to describe the probability density function of development time t. It is given by:

$$h_E(t) = \frac{\left(\frac{t_d}{m}\right)^{-m}}{(m-1)\,!} t^{m-1} e^{-m\frac{t}{t_d}} \tag{8.11}$$

Graphs of equation 8.11 are shown in figure 8.5 for values of m between 1 and 10. Order m establishes the dispersion (variance) of the Erlang function and the complexity of any process-based representation of a life stage. The mean time, $< t >$, its variance, σ^2, and the Laplace transform, H_E, of the Erlang function are given, respectively, by:

$$< t >_E = t_d \tag{8.12}$$

$$\sigma_E^2 = \frac{t_d^2}{m} \tag{8.13}$$

$$H_E(s) = \left(\frac{1}{1 + \frac{t_d}{m} s}\right)^m \tag{8.14}$$

Clearly, mean time is unaffected by order m, but dispersion decreases and complexity of representation increases with increasing m.

Bessel Description of Development. Transfer function H_I of ideal, finite delay is not realizable by a finite, process model. Another density function, here referred to as the Bessel density function, has a transfer function, denoted by H_B, that is realizable by a finite network of processes and optimally approximates H_I (Storch, 1954; Balabanian, 1958; Van Valkenburg, 1960; Smerage, 1985). With parameters m, the order of the approximation, and t_d, the delay:

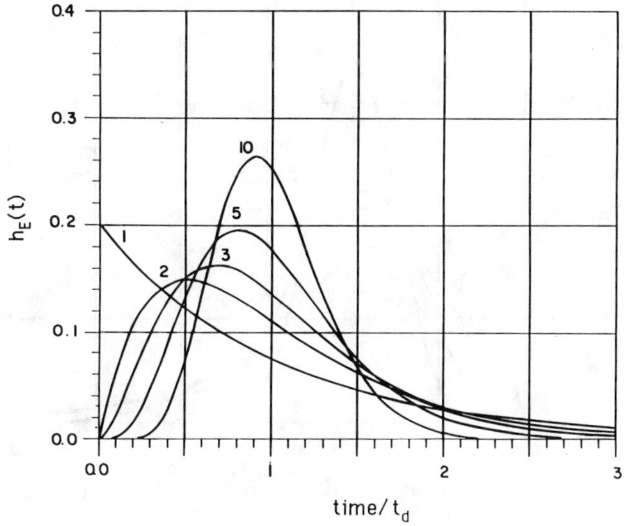

Figure 8.5–Impulse responses, h_E, of Erlang networks with time in units of mean development time t_d. The number on a curve indicates network order.

$$H_B(s) \equiv \frac{b_0}{B_m(t_d\, s)} \tag{8.15}$$

where B_m is the m*th* degree Bessel polynomial (ibid.):

$$B_m(x) = b_o + b_1\, x + \ldots + b_m\, x^m \tag{8.16}$$

Unfortunately, a closed form expression for corresponding density function h_B does not appear to exist. It may be determined by numerical methods, and graphs for several values of m between 1 and 10 are shown in figure 8.6. For large m, the Bessel density function approaches the Gaussian function (Storch, 1954). The mean time and its variance for h_B are known (Smerage, 1985).

$$< t>_B = t_d \tag{8.17}$$

$$\sigma_B^2 = \frac{t_d^2}{2m - 1} \tag{8.18}$$

Development Models

Given statistics of development time and survivorship for the various life stages and reproduction and information about other relevant processes for all

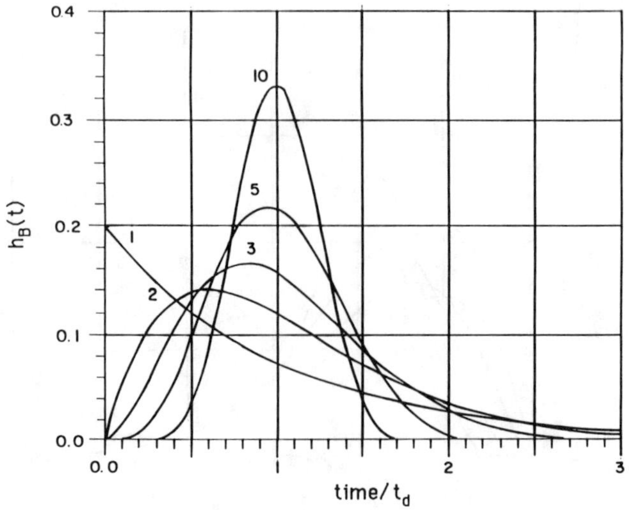

Figure 8.6–Impulse responses, h_B, of Bessel networks with time in units of mean development time t_d. The number on a curve indicates network order.

species involved, one can formulate a mathematical model corresponding to a conceptual model like that depicted in figure 8.1 of populations in an insect community. Subsequent analyses of the mathematical model by simulation and other mathematical techniques permit one to predict and interpret the dynamics of the populations in scenarios of environmental and management conditions (Tummala et al., 1975; Ruesink, 1976; Smerage et al., 1980).

Models presented here are confined to representation of the population in a single stage of the life cycle of an insect species affected only by development and physiological mortality. Convolution and network models are discussed, with emphasis on the latter. The parameters of either model are determined from statistical data on development time and survivorship in the stage, and the population occupies a region of spatial extent E (area or volume). In line with the discussion above, the probability density function of development time in the life stage will be referred to in the sequel as the impulse response, h, of the stage, and the Laplace transform of h will be denoted by H.

Concepts for building more comprehensive models incorporating reproduction, predation, parasitism, and migration build upon the life stage models presented here. Indeed, development and mortality of populations in several, successive stages or over the whole life cycle are represented by merely interconnecting models of distinct stages (ibid.).

Convolution Models. The convolution model of a single life stage in an isothermal environment is used to predict over a specified observation interval the output and population of that stage in response to a specified input. Procedurally, the output is calculated first, and then the population is calculated from that output and the specified input. With t in the observation interval, output γ_o is determined by convolving input γ_i with impulse response h of the stage (Papoulis, 1980; Director and Rohrer, 1972; Cooper and McGillem, 1967; Lathi, 1965):

$$\gamma_o(t) = \gamma_i * h = \int_{-\infty}^{t} h(t-x)\ \gamma_i(x)\, dx \tag{8.19}$$

From its value N_o at time t_o beginning the observation interval, population N at any t in the interval is calculated from:

$$N(t) = N_0 + \int_{t_0}^{t} (\gamma_i(t) - \gamma_o(t))\, dt \tag{8.20}$$

Laplace transformation of equation 8.19 yields:

$$\Gamma_o(s) = H(s)\ \Gamma_i(s) \tag{8.21}$$

and establishes a well-known fact: the transfer function of the stage, defined (ibid.) as the ratio of output to input transforms, denoted by Γ_o and Γ_i, respectively, equals Laplace transform H of h.

Equations 8.19, 8.20, and 8.21 apply as well to a sequence of life stages wherein the overall input, output, population, impulse response, and transfer function are denoted by γ_i, γ_o, N, h, and H, respectively. In addition, the overall impulse response and transfer function may be found from corresponding functions of the individual stages. Consider, for example, the three successive stages illustrated in figure 8.7, and denote their respective populations, outputs, impulse responses, and transfer functions by N_1, N_2, N_3; γ_1, γ_2, γ_3; h_1, h_2, h_3; and H_1, H_2, H_3. Note that the output of one stage equals the input to the next and $\gamma_o = \gamma_3$. With this notation, equations 8.19 and 8.20 may be employed systematically with the procedure outlined above to calculate outputs and populations over time of the individual stages. Moreover, it can be shown (ibid.) that overall impulse response h equals successive convolutions of h_1, h_2, and h_3:

$$h = h_1 * h_2 * h_3 \tag{8.22}$$

and transfer function H is given by the product:

$$H = H_1 H_2 H_3 \tag{8.23}$$

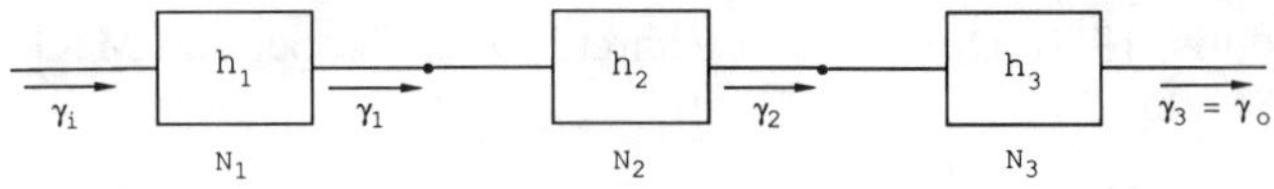

Figure 8.7–A sequence of three life stages with individual impulse responses h_1, h_2, h_3, outputs γ_1, γ_2, γ_3, populations N_1, N_2, N_3, overall input, γ_i, and overall output, $\gamma_0 \equiv \gamma_3$.

Discrete Network Models. Analyses of dynamics of populations in single and successive stages due to development with convolution models are readily implemented by computer using numerical techniques, but this approach is primarily of theoretical value. In practice, discrete network models, the mathematical models of which are systems of ordinary differential equations, are more suitable for analyses of population dynamics for partial and, in particular, comprehensive population systems that include multiple species, reproduction, predation, parasitism, and migration (Tummala et al., 1975; Ruesink, 1976; Smerage et al., 1980). Furthermore, effects of variable temperature are more readily incorporated in network models. In this section, two classes of networks and procedures for developing and analyzing them are discussed.

In theory, network models of a life stage are derivable from the probability density function of survivor development time by methods of network synthesis. In practice, mostly empirical methods have been used. Whatever the approach, the objective is to synthesize a network with an impulse response suitably approximating the impulse response, with its dispersion and delay, and survivorship of the stage represented. Whereas a convolution model provides an external or black box view of a life stage (figs. 8.1 and 8.4), a network model provides internal structure to that box. It accepts input flow, γ_i, of individuals; it retains them in the resident population until, after random delay (development time), it releases survivors at its output (γ_o). Individuals dying in the stage are transferred to a sink. Network models presented here are based on physical processes, and formal synthesis procedures exist for deriving the structure and components of such models from the system transfer function. Parameter values for the components relate to the mean and variance of the density function of development time and to survivorship in the stage.

The Components. Preliminary to presenting network models of development, the elementary components of population networks are defined (Smerage, 1985; Martens and Allen, 1969; Shearer et al., 1967). These components pertain to elementary processes defined by relations between flows and population densities, denoted by γ and n, respectively, of members of relevant populations. A terminal labeled T in the graphic symbol of any component denotes dependence of that process on temperature.

The first component, store C depicted in figure 8.8a, is used to represent a population. A store is defined by the following relation between density, n, of its population and flow, γ, of individuals incident to it.

$$C: \ \gamma = C\frac{dn}{dt} \tag{8.24}$$

Parameter C is called the capacitance of the store, and in population systems, $C = E$, the area or volume (according to the units of n) of the region occupied by the population.

Inertance I depicted in figure 8.8b pertains to inertia of flow and is defined by the following relation between flow, γ, though a path and difference in densities, $\Delta n \equiv n_1 - n_2$, between ends of the path:

$$I: \ \Delta n = I\frac{d\gamma}{dt} \tag{8.25}$$

where I is the inertance parameter.

Transformation X depicted in figure 8.8c and described by equation 8.26 below has input flow γ_1 in response to density n_1 and outputs flows γ_2 and γ_3. For transformations used in representations of life stages in population systems, γ_2 and γ_3 denote live and dead individuals, respectively; the latter go to a sink denoted by the triangle in the figure:

$$\begin{aligned} X: \ \gamma_1 &= \frac{1}{R}n_1 \\ \gamma_2 &= q\gamma_1 \\ \gamma_3 &= (1-q)\gamma_1 \end{aligned} \tag{8.26}$$

Here q denotes survivorship and R, called the input resistance of the transformation, conveys the difficulty of individuals to enter the transformation. In the case of development, that difficulty is regulated by temperature.

Erlang Networks. The probability density function of survivor development time in an age class may be approximated by Erlang function (eq. 8.11) with parameters m and t_d. A particular storage-transformation network, here called an Erlang network, realizes h_E and may be used to represent the dynamics of development in the age class. The rational function within the parenthesis of transfer function H_E (eq. 8.14) is the transfer function of the elementary store-transformation network, the first-order Erlang network, shown in figure 8.9. It realizes mean development time equal to t_d/m and variance t_d^2; it often is called the first-order or exponential delay network (Forrester, 1961). Capacitance $C = E/m$; transformation X has no mortality ($q = 1$), and its input resistance $R = t_d/E$ incorporates in X, through t_d, the dependence of development time on temperature. Time constant, τ, of the network (Chua et al., 1987; Papoulis, 1980; Director and Rohrer, 1972; Cooper

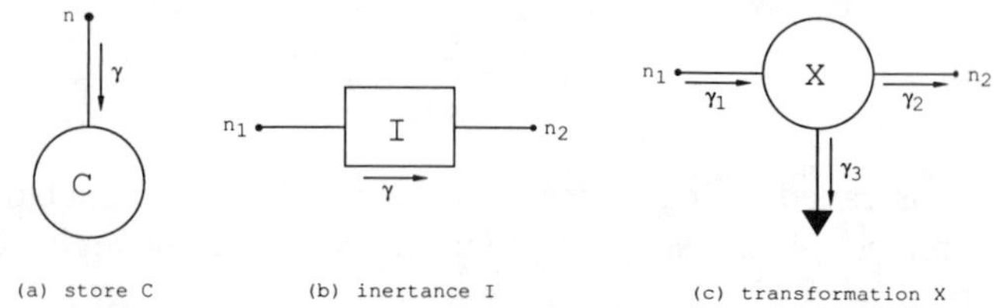

Figure 8.8–Elementary components of population networks: (a) store C; (b) inertance I; and (c) transformation X.

and McGillem, 1967; Lathi, 1965; Van Valkenburg, 1955) is given by $\tau = RC = t_d/m$.

Cascading m of the elementary networks in figure 8.9 yields the Erlang network of order m depicted in figure 8.10 which may be used to represent a life stage. Its impulse response equals the m*th* order Erlang density function with mean td and variance t_d^2/m. Capacitances, C_j, of all stores equal E/m and sum to E. Physiological mortality could have been distributed over the m substages, but, consistent with common experimental practice, that mortality has been accounted for at the end of the stage. Accordingly, q = 1 for all transformations except X_m at the output, which accounts for mortality in the whole stage, with q = survivorship of the stage. Dependence of development time on temperature is incorporated by varying t_d in the R parameters according to the t_d-temperature relationship of the species and the temperature regime.

To represent with an Erlang network the dynamics of development in a life stage of an insect species, equate parameter t_d of the network to the mean development time of the stage and set m so that the variance of network impulse response h_E, given by equation 8.13, approximates with desired accuracy the variance of development time. Then, with those values of t_d and m, build the m*th*-order Erlang network as discussed above with the structure illustrated in figure 8.10.

The synthesis procedure is illustrated here for a fifth-order Erlang network and specified t_d. In this case, m = 5 and $\sigma_E^2 = t_d^2/5$. The canonical first-order Erlang network (fig. 8.9) has C = E/5, R = t_d/E, and q = 1. Replicating the canonical network five times in series and changing q of X_5 from 1 to the survivorship of the whole stage yields the fifth-order Erlang network in figure 8.11 with transfer function:

$$H_{E5}(s) = \left(\frac{1}{1 + \frac{t_d}{5} s}\right) \tag{8.27}$$

Bessel Networks. The significance of development to population dynamics lies in the influences of time to develop and survivorship on those dynamics. Since development time relates to delay between the times survivors

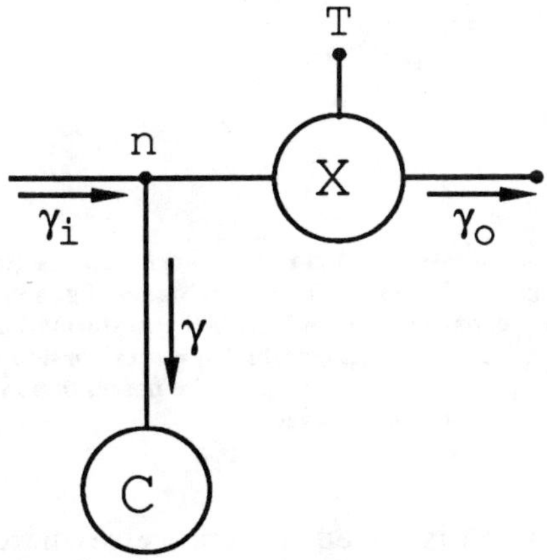

Figure 8.9–Elementary store and transformation network is first-order Erlang.

enter and exit a life stage, one could represent the dynamics of development in a life stage of an insect species with any model directly realizing mean delay and survivorship. Bessel delay networks optimally represent delay (Balabanian, 1958; Van Valkenburg, 1960), and although synthesis of Bessel networks is more involved than synthesis of Erlang networks, their performance will be shown to be better. One is led to Bessel networks in attempting to represent mean development time by ideal, finite delay and the facts that transfer function H_I of ideal delay is not realizable by a finite network and its impulse response, h_I, has no dispersion. The transfer function for the Bessel approximation, H_B, with parameters m and t_d, optimally approximates H_I (Balabanian, 1958; Van Valkenburg, 1960; Smerage, 1985); it is realizable in a finite network; and its impulse response, h_B, has dispersion controlled by m.

An m*th*-order Bessel network representation of development time in a life stage synthesized from equation 8.15 consists of stores, inertances, and an output transformation in the general configuration shown in figure 8.12 (Smerage, 1985). In contrast to an Erlang network, the stores are unequal fractions of E, but again they sum to E. The inertances also are unequal, with each proportional to t_d^2/E. Again, stage mortality is incorporated in output transformation X_o through survivorship q, and the input resistance of X_o, $R = t_d/E$, and all inertances in the network incorporate through t_d the dependence of development time on temperature.

To represent with a Bessel network the dynamics of development in a life stage of an insect species, equate parameter t_d in equation 8.15 to mean development time of the stage and set m so that variance, equation 8.18, of the network impulse response approximates with desired accuracy the variance of development time. Then, with those values of t_d and m, build an m*th*-order Bessel network with the structure illustrated in figure 8.12 using methodology

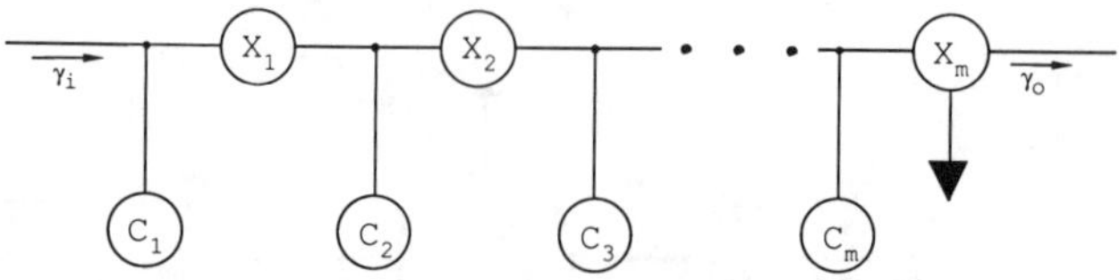

Figure 8.10–An mth-order Erlang network representation of an insect life stage consists of m elementary stages (fig. 8.9) in cascade. All stores have capacitance C_j = E/m. All transformations, X_j, have R = t_d/E and q = 1 except X_m, for which q = survivorship of the stage. Parameter E equals the spatial extent of the population in C, and t_d = mean development time of the stage.

summarized below and discussed in detail elsewhere (Smerage, 1985; Van Valkenburg, 1960; Balabanian, 1958; Storch, 1954).

Synthesis of a Bessel network is based on transfer function H_B in equations 8.15 and 8.16. Expressing B_m in equation 8.16 in terms of its even and odd parts, B_{me} and B_{mo}, respectively:

$$B_m(x) = B_{me}(x) + B_{mo}(x) \tag{8.28}$$

allows equation 8.15 to be rewritten in the form:

$$H_B(s) = \frac{\dfrac{b_0}{B_{mo}(t_d\, s)}}{1 + \dfrac{B_{me}(t_d\, s)}{B_{mo}(t_d\, s)}} \tag{8.29}$$

Extract B_{me}/B_{mo} from the denominator of equation 8.29 and label it z_{22}/R, where R is the input resistance of transformation X_o. In the terminology of network theory, z_{22} is the output impedance of the storage-inertance portion of the Bessel network in figure 8.12. Next, develop $z_{22}(s)$ in a continued fraction expansion (ibid.), where:

$$z_{22}(s) \equiv R\frac{B_{me}(t_d\, s)}{B_{mo}(t_d\, s)} = I_1\, s + \cfrac{1}{C_1\, s + \cfrac{1}{I_2\, s + \cfrac{1}{C_2\, s + \dots}}} \tag{8.30}$$

If the degree of B_{me} exceeds that of B_{mo}, remove from z_{22} by division the term I_1 s before beginning the expansion; otherwise, begin the expansion directly without I_1 s.

The final step is identification of components in the network from terms in expansion equation 8.30 and the corresponding configuration in figure 8.12.

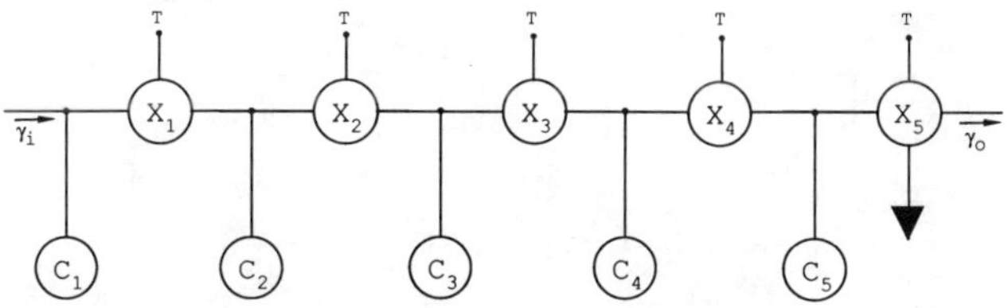

Figure 8.11–Fifth-order Erlang network representation of an insect life stage. All stores have capacitance C_j = E/5 and sum to E. All transformations, X_j, have R = t_d/E and q = 1 except X_5, for which q = survivorship of the stage.

The network will contain M stores and M or M-1 (if I_1 is absent) inertances, where M is the smallest integer ≥ m/2. Each series component in figure 8.12 is an inertance with parameter I_j identified from term I_j s in equation 8.30, and each shunt component is a store with parameter C_j identified from term C_j s in equation 8.30. The Bessel network is completed by adding at its output transformation X_o with input resistance R = t_d/E and q = survivorship of the stage. Note that synthesis of the I-C section from z_{22} starts at the output of that section and that a shunt store always appears at the input of the I-C section.

Bessel network synthesis is illustrated here for fifth-order and specified t_d. In this case, m = 5, $\sigma_B{}^2 = t_d{}^2/9$ is nearly half that of the fifth-order Erlang network, and:

$$H_B = \frac{945}{945 + 945(t_d\, s) + 420\,(t_d\, s)^2 + 105(t_d\, s)^3 + 15(t_d\, s)^4 + (t_5\, s)^5}$$

$$= \frac{\dfrac{945}{945(t_d\, s) + 105(t_d\, s)^3 + (t_d\, s)^5}}{1 + \dfrac{945 + 420(t_d\, s)^2 + 15(t_d\, s)^4}{945(t_d\, s) + 105(t_d\, s)^3 + (t_d\, s)^5}} \tag{8.31}$$

Comparing the final form of equation 8.31 with equations 8.29 and 8.30:

$$z_{22} = R\,\frac{925 + 420\,(\,t_d\, s)^2 + 15\,(\,t_d\, s)^4}{945\,(\,t_d\, s) + 105\,(\,t_d\, s)^3 + (\,t_d\, s)^5} \tag{8.32}$$

Substituting R = t_d/E and expanding z_{22} in a continued fraction:

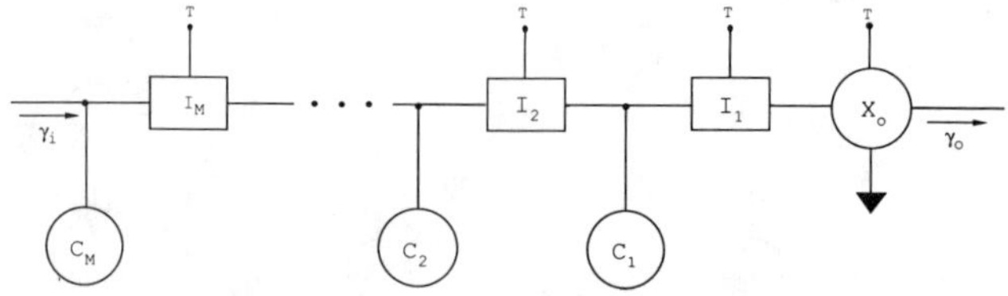

Figure 8.12–An *mth*-order Bessel network representation of an insect life stage consists of M stores, Cj, M or M – 1 (I_1 absent) inertances, I_j, and output transformation X_0. Index M is the smallest integer ≥ m/2. Transformation X_0 has R= t_d/E and q = survivorship of the stage.

$$Z_{22} = \cfrac{1}{0.067\ E\ S + \cfrac{1}{0.195\ \frac{t_d^2}{E}\ s + \cfrac{1}{0.31\ E\ s + \cfrac{1}{0.4125\ \frac{t_d^2}{E}\ s + \cfrac{1}{0.623\ E\ s}}}}} \tag{8.33}$$

Matching terms in equation 8.33 with equation 8.30 and the network architecture in figure 8.12, components of the fifth-order Bessel network shown in figure 8.13 are identified.

$C_3 =\ 0.623\ E \qquad C_2 = 0.310\ E \qquad C_1 = 0.067\ E$

$I_2 = 0.4215\ t_d^2/E \qquad I_3 = 0.195\ t_d^2/E$

X_o: R= t_d/E and q = stage survivorship

Note that only inertance parameters and R depend on mean development time, t_d, and capacitance parameters depend only on spatial extent E of the population. Dependence of development time on temperature is incorporated by varying t_d in I and R parameters according to the temperature regime and the t_d-temperature relationship of the species.

Performance of Erlang and Bessel Networks. Both Erlang and Bessel networks produce impulse responses with mean delay t_d, but they differ in the variances of those responses for specified network order m (Smerage, 1985). Comparing equations 8.13 and 8.18 for equal variances:

$$m_E = 2m_B - 1 \tag{8.34}$$

or, for identical order m:

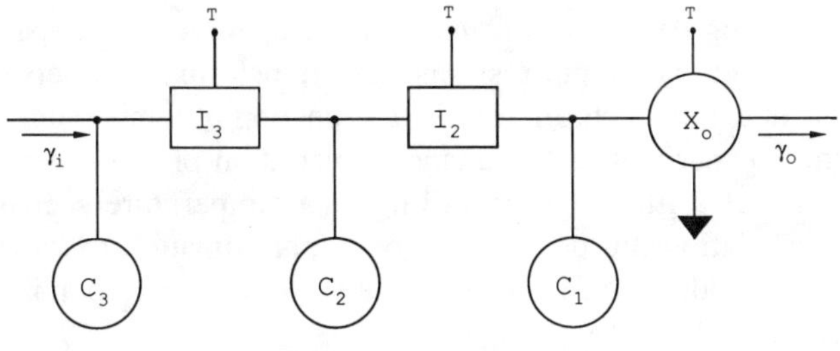

Figure 8.13–Fifth-order Bessel network representation of an insect life stage. Stores $C_1 = 0.067$ E, $C_2 = 0.310$ E, and $C_3 = 0.623$ E sum to E. Inertance $I_2 = 0.195\ t_d^2/E$ and $I_3 = 0.42\ t_d^2/E$. Transformation X_0 has $R = t_d/E$ and q = survivorship of the stage.

$$\sigma_E^2 = \left(2 - \frac{1}{m}\right)\sigma_B^2 \tag{8.35}$$

Thus, to achieve an impulse response with specified variance, equation 8.34 indicates an Erlang network up to twice the size (number of components equals number of state variables) of a Bessel network may be required. Alternatively, equation 8.35 indicates the variance of the impulse response of an Erlang network may be up to twice that of a Bessel network of identical order, m. Some feeling for the difference in performance may be had by comparing the fifth-order Erlang response in figure 8.5 (m = 5) with the fifth-order Bessel response in figure 8.6. The latter is narrower and higher in amplitude.

The transfer functions of first-order Erlang and Bessel networks are identical (compare eqs. 8.14 and 8.15 for m = 1), and they pertain to the single store, single transformation network in figure 8.9. Often called the first-order or exponential delay network (Forrester, 1961), it has been used frequently to represent individual life stages in population models. However, such use of the first-order network generally is very inappropriate, and the m = 1 curve in figure 8.5 indicates the reason: the impulse response is simple exponential decay, so immediately after an input impulse deposits a cohort in C a large portion of that cohort leaves the stage prematurely.

Variable Temperature Regimes. The discussion to this point has been for constant temperature environment, but temperature often varies widely in real systems and causes development time to vary. All concepts for representing the dynamics of the population in a life stage extend, with complexity, to variable temperature environment. Impulse response h now has additional temporal variation due to the temperature regime acting on mechanisms controlling an organism's development time. Generally, that variation will be different for different temperature regimes. Since convolution models require full specification of h over time, convolution models for life stages will be difficult to employ in all but simple cases of temperature variation.

Variable temperature regimes are more readily incorporated in network models representing development time in life stages. As previously discussed, inertance parameters and input resistances of transformations depend on mean development time, t_d, which, in turn, is the function of temperature determined in development experiments. Instantaneous variation of those parameters in a network model of a life stage according to a temperature scenario and the t_d-temperature relationship provides a good approximation to variation of the rate at which individuals move through the stage and corresponding dynamics of its population.

Applications

Convolution Models. Convolution models representing the dynamics of development in life stages of insects are useful in studies of the population dynamics of single and multiple cohorts in single, multiple, and successive life stages and over life cycles. As outlined above, they are employed using established methods of convolution (Papoulis, 1980; McGillem and Cooper, 1974; Director and Rohrer, 1972; Cooper and McGillem, 1967; Lathi, 1965). When augmented by a compatible treatment of reproduction, population dynamics throughout a life cycle over several generations could be studied with these models. However, the same information is more easily obtained from network models.

Network Models. Network models more "naturally" represent the dynamics of development in a life stage of an insect species because they are developed and interpretable in terms of processes and the biology of the life stage, while convolution models tend to be abstract and purely mathematical. Consequently, network models are easier and more intuitive to implement and interpret, especially for comprehensive systems of multiple, interacting species.

The practical choice between Erlang and Bessel network representations of life stages is based on (1) the level of complexity, as measured by m, acceptable in an analysis, and (2) how well respective impulse responses of specified orders fit relative frequency data of development times for stages to be represented. Some analysts may prefer the storage-transformation composition of Erlang networks as more natural representations of the biology than Bessel networks which, with their inertances, are artifices for representing delay efficiently in this biological application. If efficiency is important, Bessel networks may be the choice.

Steps in Producing a Network Model. Analysis by a network model of the dynamics of populations in a community of insects generally proceeds in four major steps. First, a block diagram like figure 8.1 is produced showing developmental, reproductive, interactive (predation, parasitism), and migratory processes of all relevant species in the community. Boxes depict life stages of each species with the flow of individuals out of one stage being input to the succeeding stage. Circles, like X_p, X_{r1}, and X_{r2} in figure 8.1, depict predation and reproduction, respectively, and other boxes, like M_1 and M_2, depict migratory processes. Descriptions of those processes are compatible with but beyond the scope of this discussion (Tummala et al., 1975; Ruesink, 1976; Smerage et al., 1980).

The second step is selection and substitution of a development network, like figure 8.10 or 8.12 with appropriate parameters, for each box depicting a life stage in the block diagram. Appropriate processes also are substituted for symbols depicting reproduction, predation, and migration. A network model results that represents the community in terms of stores, transformations, transport phenomena, and (with Bessel networks for life stages) inertances.

The third step is formulation of a mathematical model of the system from its network model. Several alternative methods exist (Chua et al., 1987; Athans et al., 1974; Director and Rohrer, 1972; Martens and Allen, 1969; Koenig et al., 1967), and one of those, the state variables method, may be preferred interpretively. Stores and inertances are the only processes exhibiting state, and their respective state variables are density, n, of the population in a store and flow, y, of individuals through an inertance (Smerage, 1985). The state variables mathematical model of an m*th*-order network representing development in a life stage is a set of m equations, one for each store and one for each inertance in the network expressing the rate of change of its state variable (ibid.).

Since an m*th*-order Erlang network representation of development contains only stores and transformations, its state variables pertain only to densities of populations in the m stores. In contrast, an m*th*-order Bessel network contains both stores and inertances, and its state variables pertain to densities of populations in the stores and flows through the inertances. Formulation procedures and forms of mathematical models are illustrated in the following section with state variable models for the fifth-order Erlang and Bessel networks in figures 8.11 and 8.13. Each formulation includes the first store in a network representing the succeeding stage to illustrate interstage coupling.

The fourth and final step is prediction and interpretation of dynamic behavioral properties of the real community (population system) by simulation and other mathematical analyses of the model. Concepts and methods of behavior analysis are not considered here; that large topic is well presented in the literature (Chua et al., 1987; Papoulis, 1980; Shannon, 1975; Athans et al., 1974; McGillem and Cooper, 1974; Director and Rohrer, 1972; Cooper and McGillem, 1967; Koenig et al., 1967; Lathi, 1965; Gordon, 1969). Important parts of this step are verification and validation of the model. The former seeks to establish correctness of the translation of the mathematical model into a computer program for use in computer implemented behavior analyses. The latter assesses suitability of a model for representing a specified, real system by comparing the morphologies, functions, and, in particular, dynamics in corresponding situations of the model and real systems.

Special Procedures

State variable mathematical models of Erlang and Bessel networks representing a life stage are formulated below to illustrate methods, including starting, intermediate, and final forms of equations. In each case, the primary stage, with mean development time t_d, is represented by a fifth-order network, the output of which enters a succeeding stage represented by an *Mth*-order network. Figures 8.11 and 8.13 for the primary stage are redrawn as figures 8.14 and 8.15 to show identities of variables used and the input of the succeeding stage. Each mathematical model consists of five equations for the primary stage plus an equation for input store C_M of the succeeding stage, thereby illustrating linkage of successive stages.

Erlang Network. The state variables are densities $n_1, \ldots, n_5$, of stores $C_1, \ldots, C_5$, for which $C_j = E/5 \equiv C$, and n_M of C_M. Transformations $X_1, \ldots, X_5$ have $R = t_d/E$ with t_d = delay of primary stage; X_M has input resistance R_M; $q = 1$ for all X_j except X_5, for which q = survivorship of primary stage. Beginning with the equation describing each store and successively substituting structural and component descriptions:

$$\frac{d}{dt} n_1 = \frac{1}{C} \gamma_{C1} = \frac{1}{C}[\gamma_i - \gamma_1] = \frac{1}{C}\left[\gamma_i - \frac{1}{R} n_1\right]$$

$$\frac{d}{dt} n_2 = \frac{1}{C} \gamma_{C2} = \frac{1}{C}[\gamma_1 - \gamma_2] = \frac{1}{C}\left[\frac{1}{R} n_1 - \frac{1}{R} n_2\right]$$

$$\frac{d}{dt} n_3 = \frac{1}{C} \gamma_{C3} = \frac{1}{C}[\gamma_2 - \gamma_3] = \frac{1}{C}\left[\frac{1}{R} n_2 - \frac{1}{R} n_3\right]$$

$$\frac{d}{dt} n_4 = \frac{1}{C} \gamma_{C4} = \frac{1}{C}[\gamma_3 - \gamma_4] = \frac{1}{C}\left[\frac{1}{R} n_3 - \frac{1}{R} n_4\right]$$

$$\frac{d}{dt} n_5 = \frac{1}{C} \gamma_{C5} = \frac{1}{C}[\gamma_4 - \gamma_5] = \frac{1}{C}\left[\frac{1}{R} n_4 - \frac{1}{R} n_5\right]$$

$$\frac{d}{dt} n_M = \frac{1}{C_M} \gamma_{CM} = \frac{1}{C_M}[\gamma_0 - \gamma_M] = \frac{1}{C_M}\left[\frac{q}{R} n_5 - \frac{1}{R_M} n_M\right]$$

After simple manipulations, the following, final form is obtained:

$$\frac{d}{dt} n_1 = \frac{1}{C} \gamma_i - \frac{1}{RC} n_1$$

$$\frac{d}{dt} n_2 = \frac{1}{C} n_1 - \frac{1}{RC} n_2$$

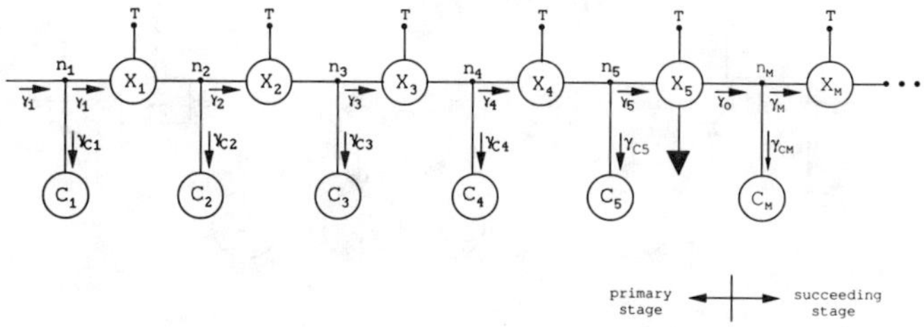

Figure 8.14–Variables used in formulating the state variables model for a fifth-order Erlang network (primary stage) and the input of an M*th*-order Erlang network representing the succeeding stage.

$$\frac{d}{dt} n_3 = \frac{1}{C} n_2 - \frac{1}{RC} n_3$$

$$\frac{d}{dt} n_4 = \frac{1}{C} n_3 - \frac{1}{RC} n_4$$

$$\frac{d}{dt} n_5 = \frac{1}{C} n_4 - \frac{1}{RC} n_5$$

$$\frac{d}{dt} n_M = \frac{q}{R\,C_M} n_5 - \frac{1}{R_M\,C_M} n_M$$

Bessel Network. The state variables are densities n_1, n_2, and n_3 of populations in C_1, C_2, and C_3 and flows γ_2 and γ_3 in I_2 and I_3 of the primary stage and n_M and γ_M in C_M and I_M of the succeeding stage. Transformation X_0 has $R = t_d/E$ and q = survivorship of primary stage. Also, $\Delta n_{I2} \equiv n_2 - n_1$ and $\Delta n_{I3} \equiv n_3 - n_2$. Beginning with the equation describing each store and each inertance and successively substituting structural and component descriptions:

$$\frac{d}{dt} n_1 = \frac{1}{C_1} \gamma_{C1} = \frac{1}{C_1}\left[\gamma_2 - \gamma_1\right] = \frac{1}{C_1}\left[\gamma_2 - \frac{1}{R} n_1\right]$$

$$\frac{d}{dt} n_2 = \frac{1}{C_2} \gamma_{C2} = \frac{1}{C_2}\left[\gamma_3 - \gamma_2\right]$$

$$\frac{d}{dt} n_3 = \frac{1}{C_3} \gamma_{C3} = \frac{1}{C_3}\left[\gamma_i - \gamma_3\right]$$

$$\frac{d}{dt} \gamma_2 = \frac{1}{I_2} \Delta n_{I2} = \frac{1}{I_2}\left[n_2 - n_1\right]$$

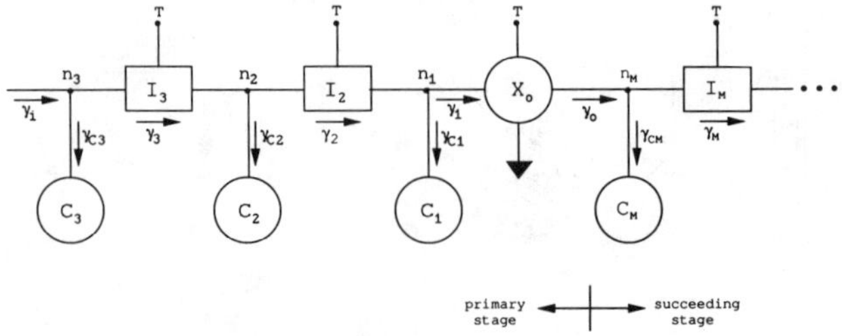

Figure 8.15–Variables used in formulating the state variables model for a fifth-order Bessel network (primary stage) and the input of an *Mth*-order Bessel network representing the succeeding stage.

$$\frac{d}{dt}\gamma_3 = \frac{1}{I_3}\Delta n_{I3} = \frac{1}{I_3}\left[n_3 - n_2\right]$$

$$\frac{d}{dt}n_M = \frac{1}{C_M}\gamma_{CM} = \frac{1}{C_M}\left[\gamma_0 - \gamma_M\right] = \frac{1}{C_M}\left[\frac{q}{R}n_1 - \gamma_M\right]$$

After simple manipulations, the following final form is obtained:

$$\frac{d}{dt}n_1 = \frac{1}{C_1}\gamma_2 - \frac{1}{R\,C_1}n_1$$

$$\frac{d}{dt}n_2 = \frac{1}{C_2}\gamma_3 - \frac{1}{C_2}\gamma_2$$

$$\frac{d}{dt}n_3 = \frac{1}{C_3}\gamma_i - \frac{1}{C_3}\gamma_3$$

$$\frac{d}{dt}\gamma_2 = \frac{1}{I_2}n_2 - \frac{1}{I_2}n_1$$

$$\frac{d}{dt}\gamma_3 = \frac{1}{I_3}n_3 - \frac{1}{I_3}n_2$$

$$\frac{d}{dt}n_M = \frac{q}{RC_M}n_1 - \frac{1}{C_M}\gamma_M$$

Conclusion

Convolution and process-based networks were shown to be suitable for representing the dynamics of development in a life stage of an insect species. This was possible through the realization that development time, physiological mortality, and their statistics are the features of development essential to population dynamics. Both types of models receive live individuals at their inputs, retain them for random intervals of time, and subsequently release survivors at their outputs in conformity with flow and residence of individuals developing in the real life stage. The basis for design of either model is for its impulse response to approximate closely the probability density function of development time and the physiological mortality in the stage.

Network models were emphasized for their relative ease of analysis and application, and the discussion focused on two types, Erlang and Bessel networks. Composed of storage, inertance, and transformation components, the size (order) of a network in either type depends on the variance of development time in the life stage represented—the smaller the variance, the larger the network required. Erlang and Bessel networks of similar order designed to produce impulse responses with identical mean development times were shown to differ in the variances of those responses, with variance of the Bessel network being smaller. Production of an impulse response with small coefficient of variance requires an Erlang network essentially twice the size of a Bessel network, a fact that may be important to computational efficiency in applications involving large models and small coefficients of variance of development times. When Erlang or Bessel networks are used to represent life stages in a comprehensive population model, formulation of the mathematical model, particularly in the state variables format, and analyses of population dynamics with that model are readily performed by well established procedures.

References

Athans, M., M. L. Dertouzos, R. N. Spann and S. J. Mason. 1974. *Systems, Networks, and Computation*. New York: McGraw-Hill Book Co.

Auslander, D. M., G. F. Oster and C. B. Huffaker. 1974. Dynamics of interacting populations. *Journal of the Franklin Institute* 297:345-76.

Balabanian, N. 1958. *Network Synthesis*. Englewood Cliffs, NJ: Prentice Hall, Inc.

Caswell, H. 1972. A simulation study of a time lag population model. *Journal of Theoretical Biology* 34:419-439.

Chua, L. O., C. A. Desoer and E. S. Kuh. 1987. *Linear and Nonlinear Circuits*. New York: McGraw-Hill Book Co.

Cooper, G. R. and C. D. McGillem. 1967. *Methods of Signal and System Analysis*. New York: Holt, Rinehart, and Winston, Inc.

Cooper, R. B. 1972. *Introduction to Queuing Theory*. New York: MacMillan.

Curry, G. L. and R. M. Feldman. 1987. *Mathematical Foundations of Population Dynamics*. College Station: Texas A&M University Press.

Curry, G. L., R. M. Feldman and P. J. H. Sharpe. 1978. Foundations of stochastic development. *Journal of Theoretical Biology* 74: 397-410.

Director, S. W. and R. A. Rohrer. 1972. *Introduction to Systems Theory*. New York: McGraw Hill Book Co.

Elmore, W. C. 1948. The transient response of damped linear networks with particular regard to wideband amplifiers. *Journal of Applied Physics* 19:55-63.

Forrester, J. W. 1961. *Industrial Dynamics*. Cambridge, MA: MIT Press.

Fredrickson, A. G., D. Ramkrishna and H. M. Tsuchiya. 1967. Statistics and dynamics of procaryotic cell populations. *Mathematical Bioscience* 1:327-374.

Gordon G. 1969. *System Simulation*. Englewood Cliffs, NJ: Prentice-Hall, Inc.

Gross, D. and C. M. Harris. 1974. *Fundamentals of Queuing Theory*. New York: John Wiley & Sons.

Gutierrez, A. P., D. W. DeMichele, Y. Wang, G. L. Curry, R. Skeith and L. G. Brown. 1980. The systems approach to research and decision making for cotton pest control In *New Technology of Pest Control*, ed. C. B. Huffaker. New York: John Wiley & Sons, Inc.

Koenig, H. E., Y. Tokad and H. K. Kesavan. 1967. *Analysis of Discrete Physical Systems*. New York: McGraw-Hill Book Co.

Langhaar, H. L. 1972. General population theory in the age-time continuum. *Journal of the Franklin Institute* 293:199-214.

Lathi, B. P. 1965. *Signals, Systems, and Communication*. New York: John Wiley & Sons, Inc.

Leslie, P. H. 1945. On the use of matrices in certain population mathematics. *Biometrika* 33:183-212.

———. 1948. Some further notes on the use of matrices in population mathematics. *Biometrika* 35:213-245.

Lotka, A. J. 1924. *Elements of Physical Biology*. Baltimore: Williams & Wilkins.

Manetsch, T. J. 1976. Time-varying distributed delays and their use in aggregative models of large systems. *IEEE Transactions of Systems, Man, Cybernetics* 6:547-553.

Martens, H. R. and D. R. Allen. 1969. *Introduction to System Theory*. Columbus, OH: C. E. Merrill Publishing Co.

McGillem, C. D. and G. R. Cooper. 1974. *Continuous and Discrete Signal and System Analysis*. New York: Holt, Rinehart & Winston, Inc.

Nicholson, A. J. and V. A. Bailey. 1935. The balance of animal populations. *Proceedings of the Zoological Society of London* 1:551-598.

Oldfield, D. G. 1966. A continuity equation for cell populations. *Bulletin of Mathematical Biophysics* 28:545-554.

Papoulis, A. 1962. *The Fourier Integral and Its Application*. New York: McGraw-Hill Book Co.

———. 1980. *Circuits and Systems*. New York: Holt, Rinehart & Winston.

———. 1984. *Probability, Random Variables, and Stochastic Processes*. New York: McGraw-Hill Book Co.

Parzen, E. 1960. *Modern Probability Theory and Its Application*. New York: John Wiley & Sons, Inc.

Pearl, R. and L. J. Reed. 1920. On the rate of growth of the population of the United States since 1790 and its mathematical representation. *Proceedings of the National Academy of Science USA* 6:275-288.

Ruesink, W. G. 1976. Status of the systems approach to pest management. *Ann. Rev. Entomology* 21:27-44.

Shannon, R. E. 1975. *System Simulation – The Art and Science*. Englewood Cliffs, NJ: Prentice-Hall.

Sharpe, P. J. H. and D. W. DeMichele. 1977. Reaction kinetics of poikilotherm development. *Journal of Theoretical Biology* 64:649-670.

Sharpe, P. J. H., G. L. Curry, D. W. DeMichele and C. L. Cole. 1977. Distribution model of organism development times. *Journal of Theoretical Biology* 66:21-38.

Shearer, J. L., A. T. Murphy and H. H. Richardson. 1967. *Introduction to System Dynamics*. Reading, MA: Addison-Wesley.

Shoemaker, C. A. 1980. The role of systems analysis in integrated pest management. In *New Technology of Pest Control*, ed. C. B. Huffaker. New York: John Wiley & Sons, Inc.

Sinko, J. W. and W. Streifer. 1967. A new model for age-size structure of a population. *Ecology* 48(6):910-918.

Smerage, G. H. 1985. Bessel delay networks for population models. *Transactions of the ASAE* 28(4):1269-1278.

Smerage, G. H., C. A. Musgrave, S. L. Poe and W. D. Eshleman. 1980. Systems analysis of Insect population dynamics. Technical Report IPM-3, Institute for Food and Agricultural Sciences, University of Florida, Gainesville.

Storch, L. 1954. Synthesis of constant-time-delay ladder networks using Bessel polynomials. *Proceedings of the IRE* 42:1666-1675.

Tummala, R. L., W. G. Ruesink and D. L. Haynes. 1975. A discrete component approach to the management of the cereal leaf beetle ecosystem. *Environmental Entomology* 4:175-86.

Van Valkenburg, M. E. 1955. *Network Analysis*. Englewood Cliffs, NJ: Prentice-Hall, Inc.

———. 1960. *Introduction to Modern Network Synthesis*. New York: John Wiley & Sons, Inc.

Verhulst, P. F. 1838. Notice sur la loi que la population suits dons son accroissement. *Corresp. Math. Phys.* 10:113-121.

Volterra, V. 1926. Variazione e fluttuazioni de numero d'individui in specie animali conviventi. *Mem. Accad. Lincei.* 2:31-113.

Wagner, T. L., H. Wu, R. M. Feldman, P. J. H. Sharpe and R. N. Coulson. 1985. Forum: *Annals of the Entomological Society of America* 78:691-704.

Weiss, G. H. 1968. Equations for the age structure of growing populations. *Bull. Mathematical Biophysics* 30:427-35.

Wilks, S. S. 1967. *Mathematical Statistics*. New York: John Wiley & Sons, Inc.

9

Modeling Insect Mortality

John L. Goodenough and Dan A. Wolfenbarger

Introduction

The relationship of insect mortality (l–Survival) to other biological insect processes is depicted in figure 9.1 (Barfield and Slansky, 1981, used by permission). The connecting lines illustrate the interdependence of the insect processes with survival of populations in the ecosystem. A multitude of references to models that include various aspects of modeling insect mortality can be found in the literature. In a favorite reference in cotton, Gutierrez and Wang (1979) discussed the appropriateness of several population models and presented a simplified diagram of an agroecosystem, complex flow diagrams of alfalfa and cotton ecosystems in terms of interactions and carbohydrate pool, the Leslie Matrix model, and differential equations of flow of a cotton ecosystem model. Mortality in terms of a loss from the defined ecosystem boundary is described in terms of sinks (losses), death, or emigration.

We include here selected brief examples of modeling insect mortality due to low and high temperature, insect quality, parasitoids, predators, and illustrate insect mortality caused by application(s) of the insecticide permethrin against a population of bollworm, *Heliothis zea,* Boddie, on cotton. No attempt is made to present an exhaustive review of available models, but only a review sufficient for illustration of the process of modeling insect mortality. Development and physiological age with 'natural' mortality associated with developmental delay are included separately by Smerage (Chapter 8). Also, increased or decreased birth rate could affect insect populations in ways similar to mortality, and is discussed by Mack and Smith in their chapter on recruitment (Chapter 10).

Procedure

The first step in modeling mortality in any biological system is to identify important mortality categories, both natural and induced. The second step, after identifying each key mortality cause and the life stages affected, a conceptual model (how you think the system works) should be formulated. This step is accomplished using best available knowledge, and should not be postponed

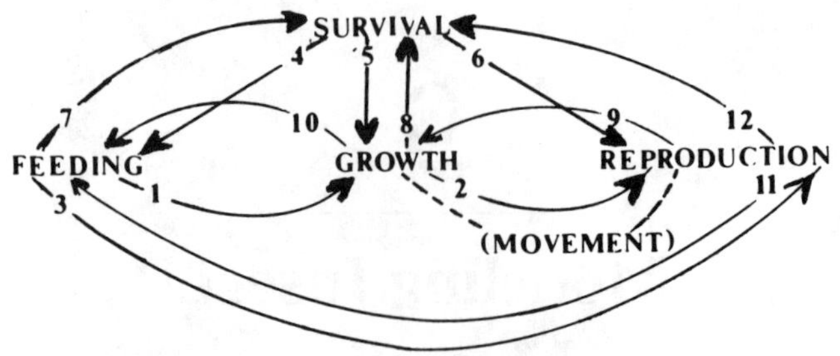

Figure 9.1–The interdependence of insect processes; note survival. (From Barfield and Slansky, 1981, used by permission.)

until the modeler feels he knows enough to develop the complete model desired. A mental picture of the system put on paper may get you started. This diagrammatic representation serves as a guide, and can be improved as more knowledge is gained. What constitutes a sufficient or appropriate degree of biological detail of each category will have to be determined for each modeling need individually. Shown in figure 9.2 is a representation of a conceptual model of a dynamic population of the parasitoid *Trichogramma pretiosum,* Riley (Goodenough and Witz, 1985). The adult *Trichogramma* population is represented as interacting with target and nontarget hosts. Various sources and sinks affect the adult population, including augmentative release of mass-reared parasites, natural and induced mortality (such as insecticide applications against pest insects), migration, and emergence from target and alternate hosts.

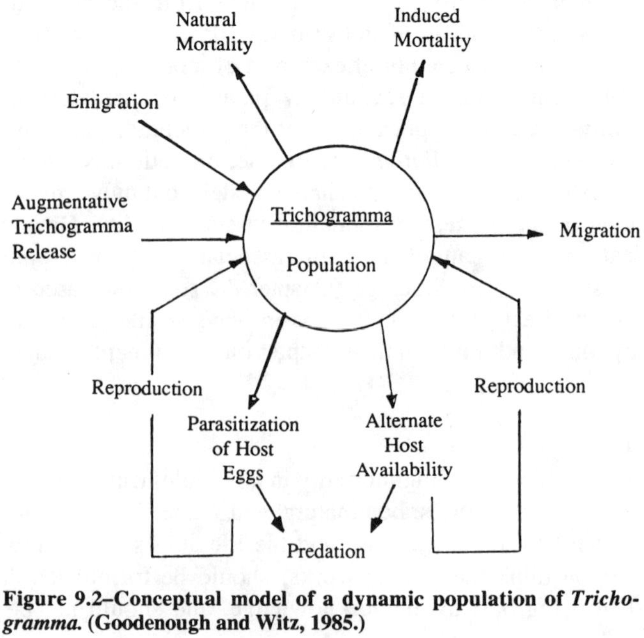

Figure 9.2–Conceptual model of a dynamic population of *Trichogramma.* (Goodenough and Witz, 1985.)

The third step, after review and revision of the beginning description of the biological system (from step 2: first try at a conceptional model), a more detailed conceptual model can and should be formulated, including available functional relationships, such as presented in figure 9.3 (Goodenough et al., 1983b). A variety of techniques might be employed to describe different functional relationships within a single model, each technique used being that which most appropriately "fits" the character of each portion of the system. At any point it may be determined that additional data are needed that are not available in the literature (i.e., it may be necessary to perform additional research).

The degree of reliability of each functional relationship should be carefully documented, both for developer and other user's reference. Reliability at this point may range from assumption to carefully documented research. However, a great deal of time and effort will be saved if this documentation is

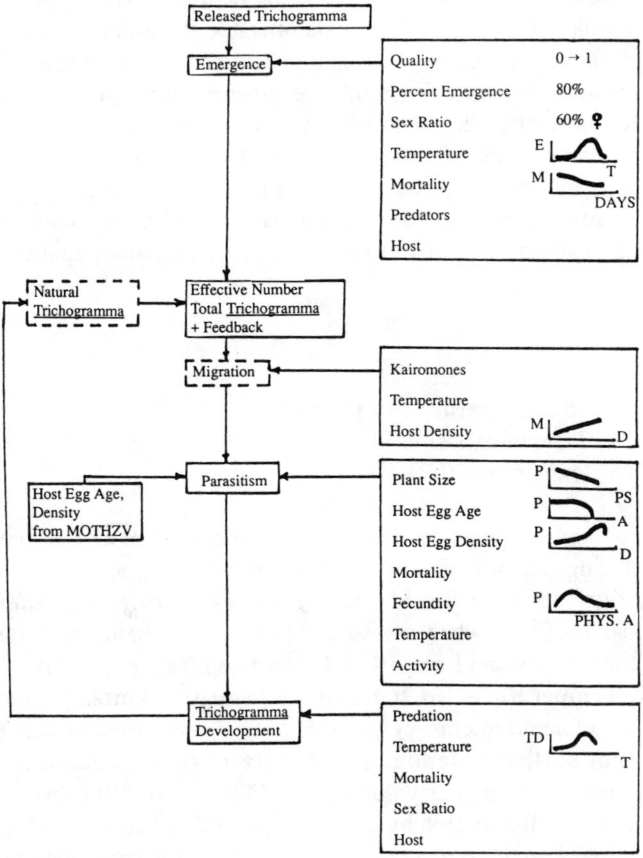

Figure 9.3–Detailed conceptual model with available functional relationships: the release of the parasitoid *Trichogramma* into a population of the pest *Heliothis zea* on cotton. (Goodenough et al., 1983b.)

done carefully and completely. For a discussion of advantages and disadvantages of 10 methods of developing conceptual models, see Chapter 4 of Jϕrgensen (1986).

The forth step, having described the problem as completely as is reasonable at this point, it is a relatively straightforward process to flowchart the computer programming required and code the computer program. We point out here that computer coding of the model is only one step in developing a model. Entering code is not modeling, but it is a necessary step in the modeling process. As additional information becomes available, the final step should be periodic review and appropriate revision of the existing model.

Examples of Insect Mortality Model

Low Temperature. Low temperature survival is important to models of overwintering survival, and some models include a low temperature survival component (Beerwinkle et al., 1978; Casagrande and Haynes, 1976; Eger et al., 1983). Beerwinkle et al. (1978) used data of stable fly, *Stomoxys calcitrans* (L)., mortality of various life stages held at constant cold temperatures to develop a model of fly mortality as a function of exposure temperature. They presented reasoning for correcting their data with Abbott's formula, for combining certain data, and for using linear regression models of the form $\ln t = b*\Theta$, where t is exposure time and Θ is exposure temperature, to develop equations of larvae and probit pupae mortalities on the natural logarithm of exposure times of 0-390 h at temperatures from 1° to –10° C. The larval and pupal model was of the form:

$$\text{Mortality} = a + * \ln t - c * \Theta \tag{9.1}$$

where

a, b, and c are constants of regression
t = Time of exposure (h)
Θ = temperature (°C)

Data on low temperature survival are also used in models to predict insect mortality during periods of cold storage, such as a quarantine treatment of perishable citrus against the Mexican fruit fly, *Anastrepha ledens* (Loew) (Anonymous, 1985), or when mass-reared insects are being held for extended periods of time (Elsey and Lam, 1978; Lopez and Morrison, 1980).

High Temperature. To simulate the increased mortality expected if an augmentative of lepidopteran eggs containing *Trichogramma* was made onto relatively hot soil, Goodenough and Witz (1985) calculated parasitoid emergence and reduced that value by predicting mortality due to high soil temperature. Upon distribution by aircraft, eggs of *Sitotroga cerealella* (Olivier) containing *Trichogramma* were scattered on both soil surface and crop (Bouse et al., 1981). The soil surface could be relatively smooth or rough and could be shaded by plants or mulch. Eggs might fall into cracks in the soil. Therefore, the eggs could be exposed to a range of soil temperatures. Soil

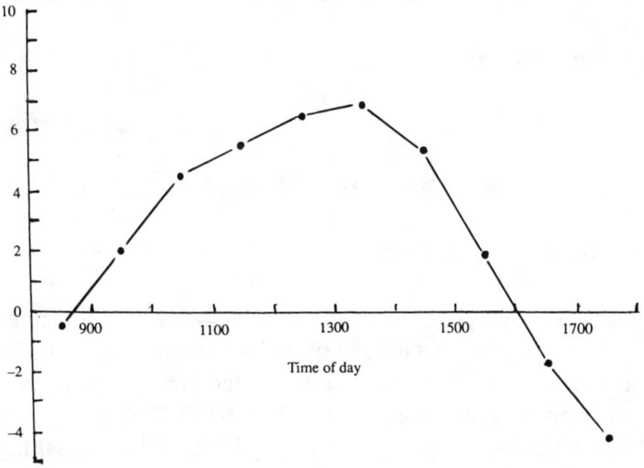

Figure 9.4–Differential between air temperature and soil temperature. Used to calculate soil temperature. (From McCann, 1985; used by permission.)

surface temperature depends on factors such as soil moisture, radiation, and soil type and slope. Goodenough and Witz (1985) estimated hourly soil temperature at 1 cm depth, assuming moderately dry, bare soil. Hourly air temperatures were calculated from max/min air temperatures using the cosine rule (see McFarland, Chapter 6). Soil temperatures were estimated by adding the differential between air temperature and soil temperature to air temperature for the respective hour, using the soil temperature model of McCann (1985) to calculate the differential between air and soil temperature for typically clear days in 1984 at College Station, Texas (fig. 9.4).

With a method to estimate hourly soil temperatures, the relationship between temperature and exposure time was required. Lopez and Morrison (1980) reported dramatic reductions in emergence and survival of *T. pretiosum* from *S. cerealella* eggs that were held for as little as 1 h at high temperatures. The eggs were held in vials in water baths at temperatures from 26.7° to 50° C. Using regression analyses, Goodenough and Witz (1985) calculated equations of fraction emerged and percent survival for three ranges of soil temperature, ST:

Fraction Emerging was:

$$52.04 * (ST - 16)/790 \qquad ST < 26$$
$$(-5613 + 405.7 * (ST) - 6.516 * (ST)^2)/790 \qquad 26 \le ST \le 40$$
$$(169 - 6 * (ST - 40))/790 \qquad ST > 40$$

The division by 790 normalized the data to the maximum emergence after 4 h at 26.7° C. ➤Y

Percent Survival was:

$$10.97 \qquad ST < 26.7$$

$$1 - 0.7 * e^{-0.263\,(40.1\,-ST)} \qquad 26.7 < ST \leq 40$$

$$0.3 - 0.03 * (ST - 40) \qquad ST \geq 40$$

The hourly soil temperature estimates were used with the emergence and survival equations to predict number of *Trichogramma* alive each hour. For instance, for a release at 1000 hours, with air temperature measured at 27° C, estimated soil temperature was 27 + 3.2 = 30.2° C (fig. 4). The number surviving was calculated from: first hour, 1000 ∗ 80% (assumed normal emergence) ∗ %EMG(ST) ∗ %SURV(ST). The latter two were from equations above for emergence and survival, respectively. For the second to fourth hour: the number emerging each hour is calculated from (number live not emerged) ∗ %EMG(ST) ∗ %SURV(ST) ∗ and added to the number already emerged. At 30.2° C, this gives, first hour:

$$1000 * 0.8 * 0.881 * (1 - 0.074) = 653 \qquad (9.2)$$

In their model, the number expected to emerge each hour after release was calculated using the poikilotherm method (Sharpe et al., 1977) as described by Goodenough et al. (1983a), and the respective rearing temperature regime (See also Smerage, Chapter 8, for calculation of developmental delay).

The revised model, TGRAMMA, coupled with the MOTHZV dynamic model of *Heliothis*, spp., allowed simulation of time of *Trichogramma* release, where mortalities would change with different max/min temperatures. A series of simulations was made to compare effects of season as reflected by the max/min temperatures and time-of-day of release on total predicted emergence and survival. Both emergence and survival were predicted to be drastically reduced for late morning releases in max/min temperature regimes 39.8/26.7 and 40.6/26.7 (temperatures which commonly occur in Texas sometime during the cotton growing season). With a 40.6/26.7 max/min temperature regime, table 9.1, predicted survival was reduced 67% and 76%, respectively, for simulated releases made at 10 A.M. and 11 A.M. compared to releases made at 8 A.M. Predictions indicated that releases should be made when soil temperature is ca. 27° to 35° C. Thus, results of the simulations could provide useful guides towards making operational decisions regarding optimal and minimal conditions under which releases should be made.

Curry et al. (1980) developed a cotton-boll weevil model that includes natural and induced mortality of the boll weevil in the field in high temperature, low moisture environments. Mortality of immature weevils is related to drying

Table 9.1. Effect of time of day on emergence and survival of released *Trichogramma* (from Goodenough & Witz, 1985)*

Hour of Release	Number of predicted adults 4 h after release with indicated max / min temperature regime†				
	32.2/23.3	35.0/23.3	35.0/26.7	39.8/26.7	40.6/26.7
8	35,449	35,544	35,525	35,299	34,862
9	35,782	35,435	34,725	32,546	28,579
10	35,105	32,058	28,796	20,365	11,295
11	32,509	23,527	19,482	9,605	8,415

* Simulated release of 50,000 *Trichogramma* each release.

† Each number is total predicted adults 4 h after a release for that particular combination of hour of release and temperature regime.

of abscised fruit, which either hang on the plant via unsevered fiber strands, or fall onto the soil. If the fruit dries beyond an amount which is estimated to allow development to the reproductive state, mortality is assessed. A developmental scale is included that allows calculation of the time to dry the minimum food mass needed for the immature weevil to develop into a viable adult as fractional drying time versus boll diameter. This allows calculation for any size susceptible fruit. Row spacing and orientation are used to calculate light penetration. Three microclimates and their relative proportions of the field are utilized: fully sunlit areas, partially shaded, and fully shaded areas. Full shade and canopy temperatures are taken as essentially ambient temperature. Jones et al. (1977) included mortality for immature stages of boll weevil as affected by temperature, but found no detailed data on the temperature affects on the longevity of adults.

Insect Quality. Insect quality and mortality are inescapably related: insects of poor quality by definition will not perform needed traits as well as will insects of high quality. Longevity is of necessity an important attribute. Insect quality of mass reared insects was described by Boller and Chambers (1977). They presented a schematic of a possible hierarchy of quality components: 1) overall quality; 2) major components: adaptability, motility, orientation to habitat, sexual activity and physiology, and reproduction; 3) traits: survival, acclimation, escape predation, dispersal, etc.; 4) measurable parameters.

For survival, measurable parameters could include longevity at various temperatures, under conditions of stress, with various diets, whether the insects are required to be mobile, amount of sexual and reproductive activity, etc. They recommended developing an initial listing of measurable parameters that is as complete as possible. Then, prioritizing should be done according to project needs and reducing the list of necessary "to dos" to items shown to have critical impact on system performance. This should be one of the important benefits of a properly structured model: the evaluation of the importance of system components to identify critical needs.

Factors influencing the quality of the parasitic wasp *Trichogramma* reared for augmentative release were summarized from several sources by Goodenough and Witz (1985). The factors included laboratory host, temperature, field host species and density, plant size or search area, behavior modifying chemicals, wind, and type, proximity, and recency of insecticide use. In their model, the factors such as overall quality, percent emergence, number of progeny/host, percent females, and rearing host and strain of *Trichogramma*, were input as decimal multiplicative constants in the form:

$$\text{Effective No. Females} = \text{No. Females (Quality} * \%\ \text{Emerg.} * \ldots)$$

This simulation does nothing towards determining actual quality, but if the assumption that these factors are multiplicative is correct, the equation is effective in estimating the impact that change in any one or more factors has on system performance (sensitivity analyses).

Although the relationship of food quality and mortality may be difficult to determine, there is ample evidence that diet has a significant impact on performance, and thus might be considered a type of mortality. Isman and Duffey (1982) found that the tomato phenolic compounds chlorogenic acid, rutin, or tomatine added to artificial diets inhibited early instar growth of *H. zea*, but not of third-fifth instar larvae. Relationship of larval weight and concentration of compound were described by linear regression.

Biological Control Agents. In a model of virus development, Johnson et al. (1982) used mortality of third instar larvae of velvetbean caterpillar, *Anticarsia gemmatalis*, to estimate nucleopoly hedrosis virus, NPV, development. They developed an *A. gemmatalis* NPV model by using the poikilotherm method of Sharpe and DeMichele (1977) for developmental rate and a quadratic function according to Sharpe et al. (1977) to distribute developmental times. Model-generated curves of percent mortality of third instar *A. gemmatalis* larvae versus days post inoculation were validated by comparison to histograms of test animal mortality.

In modeling microbial control, Cooper (1984) achieved good agreement of laboratory and field data by using feeding rate of larvae to estimate the average dose of *Bacillus thuringiensis* Berliner, B.t., spores ingested at the median application rate. From this dose the average mortality in the field for third instar larvae was estimated and compared to that obtained in the field after three days. Field degradation of B.t., spores was estimated as:

$$y = 4.22 - 1.76t \tag{9.3}$$

where y equals log (viable spore count)/leaf disk, and t equals time in days.

Unfortunately, only data of 0 and 1 days were shown. Beegle et al. (1981) presented data of B.t. on cotton leaves against *H. virescens* (F.) and *Tricoplusia ni* (Hubner) larvae. Using probit analysis, they found insecticidal half lives of about 47 h against *H. virescens* and of 37 h against *T. ni* larvae.

Parasite Mortality. Hassell (1984) discussed the reasons why the difference equation form of a model by Hassell and May (1973) was more appropriate for parasitoid modeling than a model framed in differential equations. Because parasitoids have only a single stage (adult female) searching for hosts, the omission of age structure is less important for a parasitoid model than for a predation model. Also, because parasitoids normally oviposit on or near their hosts, reproduction is directly dependent on host mortality. The equations are:

$$N_{t+1} = \tau N f\,(N_t, P_t) \tag{9.4}$$

$$P_{t+1} = N_a = N_t[1 - f\,(N_t, P_t)] \tag{9.5}$$

where

N_t, N_{t+1}, and P_t, P_{t+1} are host and parasitoid populations in successive generations

τ = net rate of increase of the hosts per generation

f = function defining the survival of hosts from parasitism

N_a = number of hosts parasitized

The model assumes that the next generation of parasitoids P_{t+1} directly results from N_a hosts parasitized.

Early modeling of *Trichogramma* was done by Knipling and McGuire (1968). They derived or assumed parameter estimates for the various efficiency factors that were identified. The model solved a differential equation for the time rate of change of an animal population. The model was used by Hartstack et al. (1976) to simulate inundative releases of this egg parasite against simulated *Heliothis* spp. populations on cotton. To generate relationships between parasite density, host density, and search area, the latter researchers used the convenient form:

$$P = 1 - e^{(-0.693 * n)/(N * SA)} \tag{9.6}$$

where

P = probability of parasitism

e = 2.7128 (natural logarithm base)

n = number of effective parasites/unit area

N = number of parasites/unit area required for 50% parasitism at SA = 1

SA = search area; for cotton, 1 at first cotton flower bud (first square); for corn, 1 when corn plants are ca. 61 cm high: for both crops, SA increases 1 unit every 10 days for ca. 50 days

This model predicts that, as the number of effective *Trichogramma*/ha (n) increases, the percent parasitism (P) increases, but at a decreasing rate (negative exponential). Parasites less efficient in searching out and parasitizing prey result in lower probability of parasitism. The probability of parasitism also decreases with increasing search area. If search area is relatively high, or parasites are ineffective in searching for and locating prey, the predicted probability of parasitism is low.

Many available papers present theoretical aspects of modeling predator/parasite population dynamics. However, few include provision for simultaneous modeling of the impact of released parasites upon a predicted pest population. In an attempt to simulate the population dynamics of *T. pretiosum* in the field, Goodenough et al. (1983b) developed a model, TGRAMMA, incorporating *T. pretiosum* development, natural mortality, and fecundity. The model calculated the number of effective female *T. pretiosum* (based on physiological age) available to parasitize a host egg population after an initial augmentative release.

The TGRAMMA model was interfaced with the modified MOTHZV model (Hartstack, 1982) and effectively modeled dramatic increases in parasitism. Numbers and quality of released *Trichogramma* (along with data on physiological age, percent emergence, and percent of population as females) are used with the host egg data to predict the percent of host eggs parasitized each day (Goodenough et al., 1983b). Host eggs not parasitized continue to develop and hatch into larvae if they are fertile. Both parasitized and nonparasitized eggs are available for predation. Parasites develop within parasitized eggs, and emerging progeny (less estimates of mortality of the adults and movement from the field) are added to naturally occurring *Trichogramma*; the total can parasitize additional new host eggs. Mortality is assessed according to physiological age, high soil or air temperature and insecticide application (Goodenough and Witz, 1985). An application of the model with expected mortality of a released parasitoid is presented above under high temperature.

Predator Mortality. The discussion of Wilson (1985) appeared particularly suited to this chapter. He discussed the impact of natural enemies in cotton and concluded that frequency of sampling is most important in estimating age-specific mortality rates due to natural enemies. When target organisms are gathered infrequently in the field and brought to the laboratory for subsequent recording of mortalities, the total mortality will be underestimated, because other mortality factors present in the field are removed in the laboratory.

Under the heading "indirect mortality estimates", Wilson (1985) included impact on pest population rate of increase, impact of pest distribution pattern, and experimental design-functional relationships as important considerations in the impact of natural enemies. Researchers might best attack the complex issue of natural enemy impact upon pest species by selecting crops having one or a few number of pests. The experimental arena should be as close to the field environment as possible. Instead of addressing each natural enemy separately, Wilson (1985) suggests it might be better to collapse all the natural enemy impact information into a common unit. The potential for and degree of control exerted by the complex of natural enemies can be quantified by comparing their growth rates under a range of natural enemy levels.

Stimac and O'Neil (1985) stated that crop/pest system models must incorporate the influence that naturally-occurring or introduced biological control agents can have on pest dynamics, if the models are going to realistically simulate crop and pest dynamics as observed in the field. They included a general equation for the mortality of insects in stage i by predators, parasites, pathogens, and other background mortality factors, respectively, assuming that the biocontrol agents act independently (from Stimac and O'Neil, 1983):

$$M_i = (m_{pred} + m_{par} + m_{path} + m_b) \tag{9.7}$$

Stimac and O'Neil (1985) insist that the key to modeling mortality by natural enemies is the attack equation, which describes the number (or percentage) of pests attacked. [Modeling the attack by parasites could be treated in a similar manner (Holling, 1959).] The attack equation is the equation which relates functionally predator and prey, and thus is central to modeling mortality induced by natural enemies. Stimac and O'Neil (1985) contend that there has been considerable study of predation and predator attack equations in agricultural systems (such as by Frazier et al. 1981), but most models of predation in agricultural systems lack the driving mechanism for predation, and thus the attack equation has no identifiable physiological, behavioral, or ecological basis.

Stimac and O'Neil (1985) presented sample models for predation of two main types. The first were physiologically based models [driven by hunger (Holling, 1965)], and incorporate the concept of handling time (Holling, 1959). The model of Gutierrez et al. (1984) illustrates the first model type. It describes predation of aphids in alfalfa:

$$M_a = (1 - e^{-b*t*M2/M1})(1 - e^{-S*M1/b}) \tag{9.8}$$

where

M_a = total prey biomass attacked
M_1 = prey biomass
M_2 = predator biomass
t = time
b = predator demand rate
S = predator discovery rate

Note that S is affected by b, and thus changes in rate of predation may be influenced by quantity or quality of prey present. The model explicitly recognizes that not all prey are of equal quality. Crop, weather, or other environmental conditions may indirectly influence the estimate of predation by changing the quantity or quality of prey available. In this type of model, searching is influenced by predator demand. Because the quality of prey is influenced by the crop nutritional state, the crop is implicitly incorporated into predator dynamics.

The second type of predator model (Stimac and O'Neal, 1985) is driven by behavior. This type of model focuses on describing the nature of searching in the system. It concentrates on leaf area that must be searched, and therefore requires estimates of leaf area and search rate. The crop is explicitly incorporated and directly influences predator search and therefore attack rates. The driving physiological mechanisms are not explicitly incorporated. Thus searching is induced through plant growth, and influences on plant growth such as fertilization and weed management directly influence predation.

The equation developed by Stimac and O'Neil (1985) describing number of *A. gemmatalis* attacked by predators in soybean is:

$$N_a = PV_L\left(C_1 e^{-C2VL} + C_3\right) \qquad (9.9)$$

where

N_a = number of attacks ➤Y
P = number of predators
V_L = number of prey (*A. gemmatalis* larvae) per unit soybean leaf area
C_1 = maximum leaf area searched when V_L is maximal
C_2 = rate of change in are searched per V_L
C_3 = minimum leaf are searched when V_L is maximal

Note that the equation contains relationships for searching and between searching and predation. Three trophic levels: crop, pest (prey), and predators are included.

Gutierrez and Baumgaertner (1984) determined coefficients for their model of female ladybird beetle, *Hippodamia convergens* Guerin, by varying the food supply (starved, intermediate, and fully fed) and measuring beetle weight and survival. They modeled the probability of survival, S, during time, dt, for individuals or for the population as:

$$0 \leq S_t = \alpha + (1-\alpha)\, N_a / (0.8bPWdt) \leq 1 \qquad (9.10)$$

where

α = 0.05
N_a = food eaten = prey
b = prey demand rate per predator
P = number of predators
W = weight of predators
dt = time period = 16 Day^o, or one day
0.8 = a constant determined by computer simulation

They found $\alpha = 0.5$ as reasonable from the experiments with starved beetles ($N_a = 0$). It is evident that determining b, N_a, etc., is not a trivial task, and other parts of their chapter describe those processes.

Insecticidal Mortality. Insecticides remain the major component of control measures (including cultural practices, resistant varieties, etc.) available for use by farmers against losses by insect pests in the Americas. Brown et al. (1979) interfaced a cotton crop model, COTCROP, a boll weevil model, BWEEV, and a *Heliothis* spp. model, HELSYS, to simulate various pest management strategies. They included the effects of insecticide applications on the boll weevil and *Heliothis* spp. populations. Adult survival after application of insecticide was assumed to be equal to a constant on days of application, and to be one on other days.

We modeled application of the synthetic pyrethroid permethrin against *H. zea*, on cotton. Needed were a means to generate estimates of pest numbers without insecticide, and adequate estimates of pest mortalities over time following insecticide applications. The MOTHZV dynamic model of Hartstack et al. (1976) was used to simulate the pest population. Data of mean numbers of male moths caught in 10 pheromone traps located in the Brazos River bottom, Burleson County, Texas, from 4 January to 8 September 1988, were used to initiate the MOTHZV model.

Estimates of mortalities of eggs, small larvae, large larvae, and adults of *H. zea* were 87%, 93%, 68%, and 83%, respectively, following an application of permethrin. These estimates were made from available data (Wolfenbarger, 1987; Wolfenbarger, unpublished data). Using the mortality coefficients and estimates of time over which the insecticide was active, daily mortality coefficients were generated using subroutine SPRAY (J. A. Witz, unpublished, described by Goodenough et al., 1983c) (table 9.2). The daily mortality estimates were multiplied by daily estimates of each pest life stage within the MOTHZV model. The method by which daily coefficients are developed is not critical, as long as suitable estimates over time are made. Percent mortality M_i on any day I is calculated using the equation:

$$M_i = A * e^{[(-(D-I)/\tau)]} \quad (9.11)$$

where

D = day of simulation
A = a coefficient calculated by binary search
τ = time constant

The coefficient, A, is calculated so that the total effect of the daily mortality over the active period of the insecticide equals the total persistence (days over which the insecticide is active against the particular life stage). It is assumed that the effect of a pesticide decays to zero in $4 * \tau$, so τ is calculated as 1/4 the persistence. (It is the product of terms j = 0 to j = 4 of $1 - A * e^{(-J/\tau)}$. (Another method would be to use standard curve-fitting techniques to obtain a

Table 9.2. Survival of *Heliothis zea* on cotton due to each application of permethrin*

Days After Application	Egg	Small Larvae†	Large Larvae†	Adult
0	0.1357	0.07324	0.3320	0.1768
1	0.4072	0.2979	0.6162	0.2080
2	0.6260	0.5518	0.7637	0.7637
3	0.7559	0.7080	0.8457	0.9287
4	0.8408	0.8096	0.8994	1.0000
5	0.8965	0.8760	0.9346	
6	0.9326	0.9189	0.9570	
7	0.9561	0.9473	0.9717	
8	0.9707	0.9648	0.9814	
9	1.0000	1.0000	1.0000	

* Developed from data of D. A. Wolfenbarger (1987), best estimates from literature, and personal experience. All parameters are permethrin, a synthetic pyrethroid, against targets of 100% *H. zea.*

† LIFE STAGES: Small Larvae, Stadia 1-3; Large Larvae, Stadia 4-6.

best-fit of mortality or survival over time after an application, and use these values of mortality or survival to reduce daily estimates of pest numbers).

For our pest population simulations, inputs to the MOTHZV model were daily max/min temperatures, mean numbers of moths as described above, and expected yield for the field of 892 kg/ha (2 bales/ac). The model was initialized for fruiting patterns of corn and cotton using means from four fields (Latheef et al., unpublished data). For corn, mean first day of silk occurred D145 (day-of-year 145 in 1988 was 24 May) and peak silking occurred about D165. These parameters are important to synchronization of the oviposition pattern (Lopez et al., 1978). For cotton, mean day of first square was D160 and mean day of first open boll D216. From these cotton fruiting data, a series of initiating simulations was made, and a cotton fruiting rate factor of 0.105 was determined. Different cotton cultivars may require different factors. Using this factor required the model generate a fruiting pattern that closely approximated that of the Latheef field data.

The following modeling procedure was used to simulate populations of *H. zea* in corn, using parameters described above, with 76,200 plants/ha and the cannibalism option (Y). The prediction of migration of *H. zea* moths that developed from corn and were assumed to disperse into cotton was used to generate the pest population moving into cotton. Insecticide was applied to cotton (but not to corn) when pest larvae estimates exceeded 2500/ha or when the damaged boll estimate exceeded 2500/ha. The prediction of larval numbers was reduced dramatically with one (D188) or two (D188, D196) applications of permethrin (fig. 9.5). Then we compared numbers of fruit under various regimes to decide whether an application would be profitable. Small boll numbers increased markedly with one or two applications (fig. 9.6), but little potential remained to increase boll numbers (and thus yield) if additional applications were made [difference between regimes labeled "with insects—no applications" and "two applications" (fig. 9.6)].

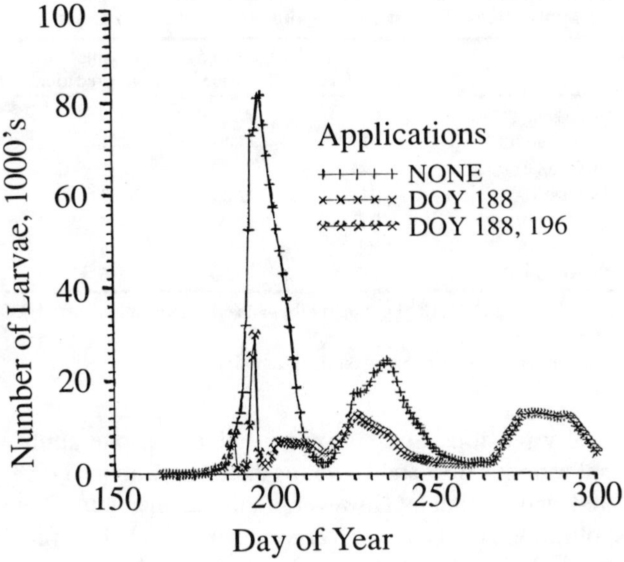

Figure 9.5–Reduction in *Heliothis zea* larval numbers predicted with one or two application of permethrin.

Estimated yields under the various management regimes were used with cotton price of \$1.10/kg to determine if a simulated application of insecticide

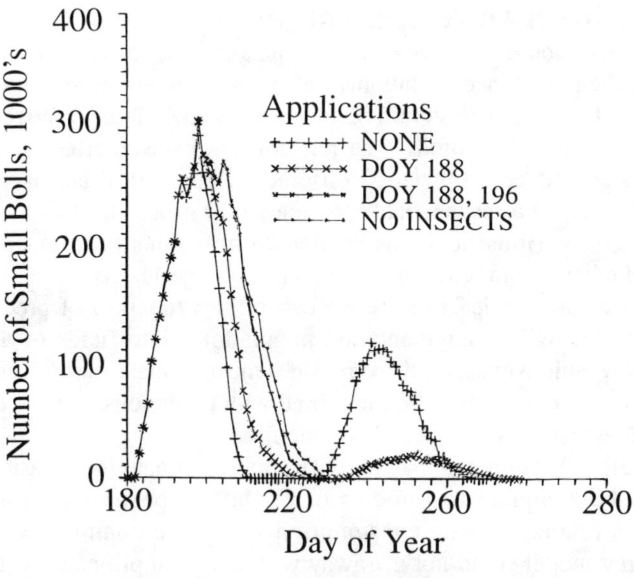

Figure 9.6–Predicted increase in small boll numbers with one or two applications of permethrin, and potential numbers of this fruit with all pests removed.

Table 9.3. Estimated yields, net return, and control decision for regimes of zero to four applications of permethrin against *Heliothis zea* on cotton

Number†	Regime	kg Lint / ha	Increased Revenue	Regime Compared to	Regime Recommended
1	No insects	1109			
2	Insects, no IC	675	-477.40	1	No
3	1X D188	740	71.50	2	Yes
4	2X D188 196	942	222.20	3	Yes
5	3X D188 196 209	962	22.00	4	No†
6	3X D188 196 223	1059	106.70	4	Yes†
7	4X No. 6 + D230	1077	19.80	6	No

* Based on cotton lint price of $1.10 / kg ($0.50 / lb) and aerial application cost of $30 / application of permethrin / ha.

† Recommend obtaining further data based on field sampling.

was economically feasible. Summarized in table 9.3, one application made D188, two applications D188 and D196, or three applications D188, D196, and D223 were predicted profitable. However, a third application D209 (strategy 5) or a fourth application D230 (strategy 7) were not predicted profitable. Thus the control recommendation was "NO" for the last application (i.e., an expected unprofitable) in both of those regimes. About 30 simulations was required for the total of the regimes shown. Most of the time using this procedure was spent obtaining initiation data or in planning strategy. Little time was needed for computer computations; each simulation required less than one minute on our COMPAQ 360 machine.

Future Model Development

Routines should simulate effects of parasite quality or field performance because of their influence on numbers of parasites required to control the pest population. These routines should include quantitative data on movement, field search rate and handling times. The reduction in parasite effectiveness due to insecticides should be estimated (House et al., 1980; Bull and House, 1983). Also needed are data on the field efficiency of various strains of mass-reared parasitoids on various hosts, as related to host density, plant phenology, presence of other natural enemies and other (nontarget) hosts.

Routines are needed to simulate capabilities (quality) of predators, both naturally occurring and augmentation populations, into fields of a particular commodity. Predictiveness of these routines must be maximized. Then proof of their efficaciousness needs to be monitored with reflections on their quality and ability to affect pest species of that commodity.

Hassell (1984) argued that simplifications adequate for parasitoid models would not be adequate for models of predation, because reproduction of predators is related to prey consumption in a more complex way. "A true predator-prey model could not get away with these simplifications (lack of age structure and direct dependence of reproduction on host mortality) and would need to include at least some provision for age-dependent differences in searching efficiency and a more realistic treatment of predator reproduction."

It seems most appropriate for our purpose here to include the somewhat critical but very significant comments of Stimac and O'Neil (1985) concerning needs for modeling predation. We recommend that persons aspiring to develop appropriate models of natural enemies consider these comments carefully. The comments are summarized nearly verbatim. "In agricultural systems, there have been few models developed that capture the essence of predation. In most crop/pest models, natural enemy effects are often omitted or are treated parametrically. Few models incorporate relevant predator behavior, or the impact of system level changes. Where predation has been studied, with the objective of mathematically describing the process, new insights into predation in agricultural systems have followed."

"Although alternative habitats and alternative prey items have been demonstrated to be vitally important to predation dynamics, few crop/pest models have alternative habitats or prey item components. Although predator searching behavior has long been identified to be one of the key elements in predation dynamics, few experimental studies have focused on searching behavior. In agricultural systems there has been little progress made in incorporating our empirical understanding of the process of predation into models of natural enemy dynamics. The net result is that we have few adequate models of predation or parasitism of pest in agricultural crop systems and have yet to fully utilize the power of crop/pest models for the study of predation in agricultural systems."

Models should include the capability to incorporate biological agents against real or theoretical pest populations, including effects of insecticides and nontarget organisms. Most biological agents have restrictions, the most important of which is their short biological life. The critical restrictions emphasize the need for precision in their time of application, lest they be ineffective. Often biological agents require high uneconomical use rates under certain conditions to obtain acceptable control of target pests. The diversity of application parameters and efficacy indicates that appropriate prediction by models will be of major importance. This trend can be expected to continue as new novel biological agents, from biotechnological research become available. As this trend continues, prediction capability will be of major importance to sound operation of agricultural management programs.

Rowell and Markham (1984) described a model that incorporates the mortality risk in one group relative to that in another (control) group. By basing their model on numbers still alive at each time in contrast to fitting survival curves to observed proportions surviving over time, they calculated a 'risk' of death for each interval as the number dying expressed as a proportion of the number alive at the beginning of the interval.

Having accurate estimates of insecticidal mortality for prediction of *Heliothis* spp., population dynamics with the MOTHZV model was shown by Goodenough et al. (1984) to be critical if reliable predictions are to be made. They stated, "Continued efforts are needed to update the insecticidal parameter estimates as new information and knowledge of field effects of pesticides becomes available. The new coefficients should include consideration of effects due to species differences, possible development of insecticidal resistance, and rate of application of insecticide".

Knipling (1980) emphasized the need for realistic estimates of fecundity, overwintering survival and migration in the evaluation of large-area control programs.

Including more accurate estimates of mortality in insect population models will lead to more accurate prediction of insect population behavior and dynamics. The key may lie in determining which aspects of any particular biological model are worthy of expenditure of time, talents and money. Some practical aspects of the problems of modeling insect mortality are presented in this Chapter. As is the usual case, the decision of which method to use for what problem is of necessity left to the reader.

Acknowledgment. Thanks to the authors who willingly allowed us to use figures from their papers to illustrate various techniques of modeling insect mortality. Thanks to J. Allen and G. Smerage for critical reviews of the manuscript.

In cooperation with the Texas Agricultural Experiment Station, Texas A&M University, College Station, Texas. Mention of a trade name or a proprietary product is for specific information only and does not constitute a guarantee or warranty of the product by the U.S. Department of Agriculture or the Texas Agricultural Experiment Station and does not imply endorsement of the product over other products not mentioned. Not for use in sales promotion or advertising which expresses or implies endorsement of the product by the U. S. Department of Agriculture. Contribution from USDA-ARS in cooperation with the Texas Agricultural Experiment Station.

References

Anonymous. 1985. Plant quarantine manual. Section VI T-107. USDA, Animal and Plant Health Inspection Service.

Barfield, C. S. and F. Slansky. 1981. Response of insect populations to dynamic crop environments. ASAE Paper No. 81-4068. St. Joseph, MI: ASAE.

Beegle, C. C., H. T. Dulmage, D. A. Wolfenbarger and Ed Martinez. 1981. Persistence of *Bacillus thuringiensis* Berliner insecticidal activity on cotton foliage. *Environmental Entomology* 10:400-401.

Beerwinkle, K. R., I. L. Berry and S. E. Kunz. 1978. Prediction models for mortality of immature stable flies caused by cold temperature. *Environmental Entomology* 7:273-277.

Boller, E. F. and D. L. Chambers. 1977. Concepts and approaches. In *Quality Control. An Idea Book for Fruit Fly Workers,* eds. E. F. Boller and D. L. Chambers, 4-13. Bulletin Int'l. Org. for Bio. Control of Noxious Animals and Plants. West Palaearctic Regional Section (SROP/WPRS Bulletin 1977/5).

Bouse, L. F., J. B. Carlton and R. K. Morrison. 1981. Aerial application of insect egg parasites. *Transactions of the ASAE* 24(5):1093-1098.

Brown, L. G., R. W. McClendon and J. W. Jones. 1979. Computer simulation of the interaction between the cotton crop and insect pests. *Transactions of the ASAE* 22(4):771-774.

Bull, D. L. and V. S. House. 1983. Effects of different insecticides on parasitism of host eggs by *Trichogramma* pretiosum Riley. *Southwest. Entomology* 8:46-53.

Casagrande, R. A. and D. L. Haynes. 1976. A predictive model for cereal leaf beetle mortality from sub-freezing temperatures. *Environmental Entomology* 5:761-769.

Cooper, D. J. 1984. The application of a model to achieve predicted mortality in a field trial using *Bacillus thuringiensis* to control *Heliothis punctiger. Entomology Exp. Applications* 36:253-259.

Curry, G. L., P. J. H. Sharpe, D. W. DeMichele and J. R. Cate. 1980. Towards a management model of the cotton-boll weevil ecosystem. *Journal of Environmental Management* 11:187-223.

Eger, J., Jr., W. L. Sterling and A. W. Hartstack, Jr. 1983. Winter survival of *Heliothis virescens* and *Heliothis zea* (Lepidoptera: Noctuidae) in College Station, Texas. *Environmental Entomology* 12:970-975.

Elsey, K. D. and J. J. Lam, Jr. 1978. *Jalysus spinosus*: Instantaneous rate of population growth at different temperatures and factors influencing the success of storage. *Entomological Society of America* 71:322-324.

Frazier, B. D., N. Gilbert, V. Nealis and R. D. Raworth. 1981. Control of aphid density by a complex of predators. *Canadian Entomology* 113:1035-1041.

Goodenough, J. L., A. W. Hartstack and E. G. King. 1983a. Developmental models for *Trichogramma pretiosum* (Hymenoptera: Trichogrammatidae) reared on four hosts. *Journal Econ. Entomology* 76:1095-1102.

Goodenough, J. L., A. W. Hartstack and K. A. Pieprzica. 1983b. Dynamic model of a pest insect parasite. ASAE Paper No. 83-1005. St. Joseph, MI: ASAE.

Goodenough, J. L., A. W. Hartstack and J. W. Witz. 1983c. Simulation of early season spray strategies and effects on beneficial insects of cotton. *Proceedings Beltwide Cotton Prod. Research Conference*, 159-167.

Goodenough, J. L., J. A. Witz and D. A. Wolfenbarger. 1984. Insecticides and beneficial insects: A new look. ASAE Paper No. 84-1006. St. Joseph, MI: ASAE.

Goodenough, J. L. and J. A. Witz. 1985. Modeling augmentative releases of *Trichogramma pretiosum*. In *Biological Control of Bollworm and Tobacco Budworm in Cotton by Augmentative Releases of* Trichogramma, eds. E. G. King, D. L. Bull, L. F. Bouse, and J. R. Phillips, 169-189. Supplement 8. Southwestern Entomology.

Goodenough, J. L. and D. R. Rummel. 1988. MOTHZV model tested as an insecticide application aid against *Heliothis zea* on cotton in the Texas High Plains in 1987. *Proceedings Beltwide Cotton Prod. Research Conference*, 260-264.

Gutierrez, A. P. and Y. Wang. 1979. Applied population ecology: Models for crop production and pest management. In *Pest Management,* proceedings of an International Conference, eds. G. A. Norton and C. S. Holling, 25-29 October 1976. Pergamon Press.

Gutierrez, A. P. and J. U. Baumgaertner. 1984. Modeling predation. In *Pest and Pathogen Control: Strategic, Tactical, and Policy Models*, ed. G. R. Conway, 93-108. New York: John Wiley & Sons.

Gutierrez, A. P., J. U. Baumgaertner and C. G. Summers. 1984. Multiltrophic models of predator-prey energetics. *Canadian Entomologist* 116:923-963.

Hartstack, A. W. 1982. Modeling and forecasting *Heliothis* populations in the U.S.A. In *Proceedings of the International Workshop on Heliothis Management*, 51-60. 15-20 November 1981. Patanchoru, A.P., India: International Crops Res. Institute for the Semi-Arid Tropics.

Hartstack, A. W., Jr., J. A. Witz, J. P. Hollingsworth, R. L. Ridgway and J. D. Lopez. 1976. MOTHZV-2: A computer simulation of *Heliothis zea* and *Heliothis virescens* population dynamics. U.S. Department of Agriculture-ARS-S-127 (Users Manual).

Hassell, M. P. 1984. Host-parasitoid models and biological control. In *Pest and Pathogen Control Strategic, Tactical, and Policy Models*, ed. G. R. Conway, 73-92. New York: John Wiley & Sons.

Hassell, M. P. and R. M. May. 1973. Stability in insect host-parasite models. *Journal Animal Ecology* 42:693-736.

Holling, C. S. 1959. Some characteristics of simple types of predation and parasitism. *Canadian Entomologist* 91:385-398.

Holling, C. S. 1965. The functional response of predators to prey density and its role in mimicry and population regulation. *Mem. Entomological Society of Canada* 45:3-60.

House, V. S., J. R. Ables, R. K. Morrison and D. L. Bull. 1980. Effect of diflubenzuron formulations on the egg parasite *Trichogramma pretiosum. Southwestern Entomology* 5:133-138.

Isman, M. B. and S. S. Duffey. 1982. Toxicity of tomato phenolic compounds to the fruitworm, *Heliothis zea. Entomological Experiments & Applications* (Amsterdam) 31:370-376.

Johnson, D. W., D. B. Boucias, C. S. Barfield and G. E. Allen. 1982. A temperature-dependent development model for a nucleopolyhedrosis virus of the velvetbean caterpillar, *Anticarsia gemmatalis* (Lepidoptera: Noctuidae). *Journal Invertebrate Pathology* 40:292-298.

Jones, J. W., H. D. Bowen, R. E. Stinner, J. R. Bradley, Jr. and J. S. Bacheler. 1977. Simulation of boll weevil population as influenced by weather, crop status and management practices. *Transactions of the ASAE* 20(1):121-125, 131.

Jφrgensen, S.E. 1986. *Fundamentals of Ecological Modeling*. New York: Elsevier.

Knipling E. F. 1980. Regional management of the fall armyworm — A realistic approach? *Florida Entomology* 63:468-480.

Knipling, E. F. and J. U. McGuire, Jr. 1968. Population models to appraise the limitations and potentialities of *Trichogramma* in managing host insect populations. U.S. Department of Agriculture, Bulletin No. 1387.

Lopez, J. D. and R. K. Morrison. 1980. Effects of high temperatures on *Trichogramma pretiosum* programmed for field release. *Journal Econ. Entomology* 73:667-670.

Lopez, J. D., A. W. Hartstack, Jr., J. A. Witz and J. P. Hollingsworth. 1978. *Heliothis zea*: oviposition on corn and sorghum in relation to host phenology. *Southwestern Entomology* 2:158-164.

McCann, I. A. 1985. An area-wide soil temperature model for predicting *Heliothis zea* and *Heliothis virescens*. Ph.D. diss. Texas A&M University, College Station.

Rowell, J. G. and P. G. Markham. 1984. Modeling mortality risks for treated insects relative to controls. *Annals of Applied Biology* 105:15-27.

Sharpe, P. J. H., G. L. Curry, D. W. DeMichele and C. L. Cole. 1977. Distribution model of organism development times. *Journal of Theoretical Biology* 66:21-38.

Sharpe, P. J. H. and D. W. DeMichele. 1977. Reaction kinetics of poikilotherm development. *Journal Theoretical Biology* 64:649-670.

Stimac, J. L. and R. J. O'Neal. 1983. Modeling the impact of natural enemies on insect pests in soybeans. In Natural enemies of arthropod pests in soybeans, ed. H. N. Pitre, 78-87. Southern Cooperative Series Bulletin No. 285.

Stimac, J. L. and R. J. O'Neal. 1985. Integrating influences of natural enemies into models of crop/pest systems. In *Biological Control in Agricultural IPM Systems*, eds. M. J. Hoy and D. C. Herzog, 323-344. New York: Academic Press, Inc.

Wilson, L. T. 1985. Estimating the abundance and impact of arthropod natural enemies in IPM systems. In *Biological Control in Agricultural IPM Systems*, eds. M. J. Hoy and D. C. Herzog, 303-322. New York: Academic Press, Inc.

Wolfenbarger, D. A., et al. 1987. Efficacy of chemical insecticides and insect-growth regulators against *Heliothis* species. In *Theory and Tactics of Heliothis Population Management II. Insecticidal and Insect Growth Regulator Control*, eds. J. C. Schneider et al., 16-31. Southern Cooperative Series Bulletin No. 329.

10

Modeling Insect Recruitment

Timothy P. Mack and James W. Smith, Jr.

Introduction

An insect population is a dynamic entity with its oscillations dictated by rates of natality, mortality, immigration, and emigration. Population increases are the direct result of the addition of new individuals by either natality and/or immigration, which is collectively called recruitment. Accurate estimates of recruitment are essential for modeling population dynamics and forecasting increases in population density. For the purpose of this discussion, recruitment is defined as the addition of new individuals by births. Thus, this chapter focuses on insect oviposition, the factors that influence it, and methods of modeling oviposition. Methods of modeling immigration are discussed in Chapter 11 by Westbrook.

Oviposition is an essential component of insect population dynamics, because no new individuals would be added to a population without oviposition or immigration. Oviposition assumes critical importance in multi-generation simulation models where future populations of nymphs or larvae are predicted from the extant adult population. For insects that are agricultural pests, oviposition may be a very important factor in management planning (Regan et al., 1979; Shoemaker et al., 1986). Biologically realistic algorithms are needed in models to accurately estimate oviposition. The degree of realism required is determined by the degree of accuracy deemed acceptable in the model, and by the objectives of the model. Thoughtful selection of model objectives is a basic step in the modeling process (Gold, 1975), and greatly affects the accuracy needed in the oviposition submodel(s). For example, a model designed to assess control strategies for a particular insect (e.g., Greever and Georghiou, 1979) may need little information relative to oviposition rates, while a research model designed to simulate multi-generation population development may demand an exacting and detailed oviposition submodel, such as the boll weevil model (Jones et al., 1975).

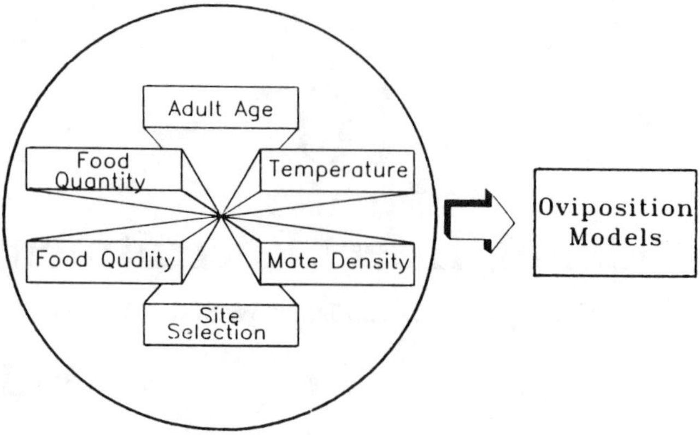

Figure 10.1–Factors commonly affecting oviposition in most insects.

Factors Affecting Oviposition

Any factor that directly or indirectly affects an insect may affect oviposition and egg viability. The relative importance of each factor will vary among species and environmental situations. However, several factors commonly affect oviposition in most insects (fig. 10.1). These factors, while not of critical importance for modeling every insect, are important in a biological sense and should be considered for inclusion in oviposition models.

Adult Age. In general, higher organisms must reach an age of sexual maturity before reproduction. Many of these organisms achieve a peak of reproductive activity shortly after sexual maturity. After this peak, reproductive activity generally declines (Spain, 1982). The curve that describes this activity is generally asymmetrical, and is skewed towards older individuals (fig. 10.2). A gamma distribution has been commonly employed to describe asymmetrical oviposition rate. Both a gamma distribution and a sixth-order polynomial described the oviposition rate of the lesser cornstalk borer, *Elasmopalpus*

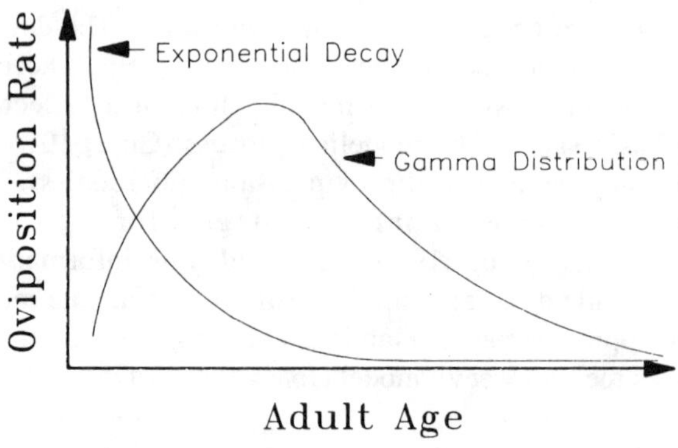

Figure 10.2–Examples of variation in oviposition rate with adult age. (After Spain, 1982.)

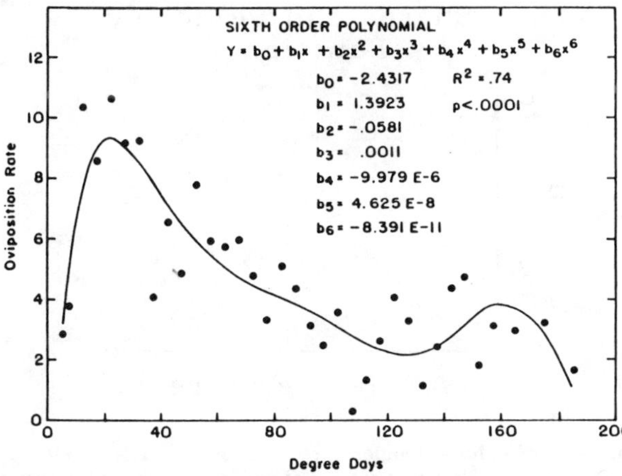

Figure 10.3–Fluctuations in oviposition rate for adult female *Elasmopalpus lignosellus* with adult age as measured in degree-days. (From Mack et al., 1987.)

lignosellus (Zeller); however, the latter function better described a small secondary peak in oviposition prior to the death of the adults (fig. 10.3) (Mack et al., 1987). Bimodal (e.g., Gutierrez et al., 1976) and trimodal curves (Seawright et al., 1977) have been employed to describe the relationship between adult age and oviposition rate. Linear equations (Hughes and Sands, 1979), square root functions (e.g., Brown et al., 1978), and constant functions (e.g., Mack and Smilowitz, 1982) also have been employed for oviposition rate calculations.

Temperature. Oviposition rate, like other rate processes in poikilothermic organisms, varies with temperature. We define the maximum oviposition rate per day, O_{max}, as the oviposition rate at a given temperature for a poikilothermic organism that is not limited by food quantity or quality, or by mate availability. O_{max} should be described by a nonlinear temperature function such as that developed by Sharpe and DeMichele (1977) for poikilothermic development rates (fig. 10.4), assuming one rate-limiting step in the oogenesis process. The Sharpe and DeMichele equation (1977) is:

$$R = \frac{\{T \times \exp[(\beta - \tau)/T]/r\}}{\{1 + \exp[(\delta - \Phi)/T]/r\} + \{\exp[(\Omega - \Phi)/T]/r\}} \tag{10.1}$$

where

R	=	rate (in our discussion, O_{max})
T	=	temperature in ° K
r	=	a physical constant independent of the organism
τ, β, Ω, Φ, and δ	=	parameters which reflect the individual thermodynamic characteristics of the organism's enzyme system

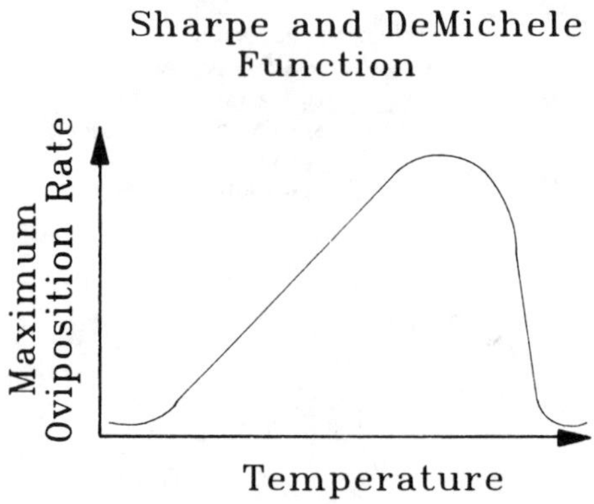

Figure 10.4–Nonlinear temperature function typically employed to describe insect development rates. The maximum oviposition rate per day for many insects (see text) should also approximate this shape.

The diversity of methods employed by authors to quantify temperature effects on oviposition rate parallels the methods used for development rate-temperature studies. Most popular among the linear methods for both development and oviposition is the degree-day approach. Degree-days have been employed to describe the fluctuations in oviposition rate with temperature (Hardman, 1978; Hughes and Sands, 1979; Gutierrez et al., 1981; Flynn et al., 1986). The use of degree-days for oviposition can have a greater effect than development within a generation, because oviposition *adds new individuals* to the population, whereas development only increases or decreases the rate at which individuals mature or age. The degree-day assumption will be valid only within the linear portion of equation 10.1 (Sharpe and DeMichele, 1977), and extrapolation beyond the optimal temperature for oviposition could lead to major errors in the number of new individuals accumulated in a model. Mortality is typically large when either the upper or the lower thermal limit of oviposition is approached. Most insects have adapted mechanisms to avoid these extremes, however.

An example of a curvilinear oviposition function can be found in Wilkerson et al. (1986), where a polynomial function described the effects of temperature on oviposition of velvetbean caterpillars adults, *Anticarsia gemmatalis* Hùbner. Mason and Mack (1984) described the relationship of temperature and adult age to *Pseudoplusia includens* (Walker) oviposition rate with a two-variable polynomial (fig. 10.5), where two peaks in oviposition occurred. The major peak was the typical peak commonly found for many insects. The minor peak occurred shortly before the death of the adult females, and may represent the insect's last attempt at reproduction.

Very few models have used nonlinear functions to describe oviposition rate changes with temperature. Toole et al. (1984) used an analytic equation (Logan et al., 1976) similar in form to equation 10.1 for their oviposition rate

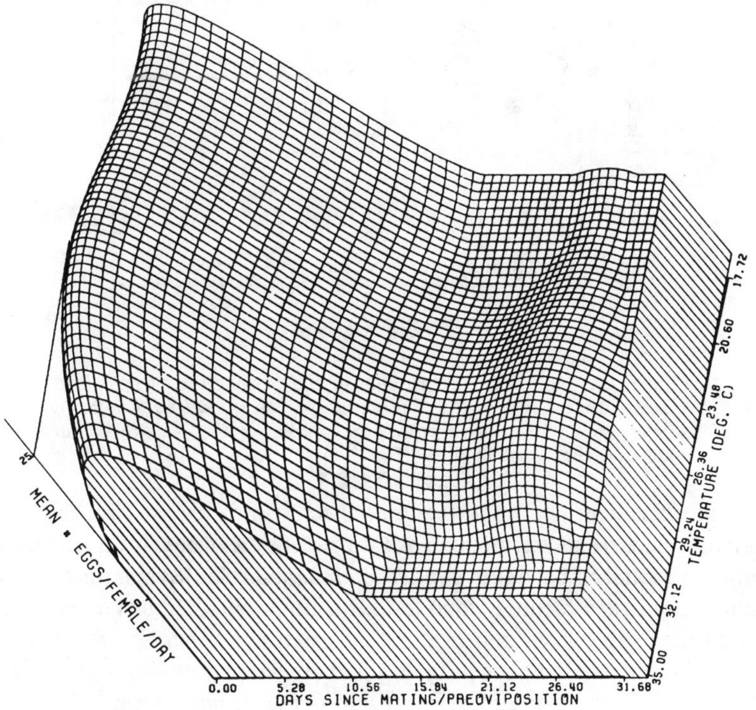

Figure 10.5–Two-variable polynomial response surface describing the relationship between temperature and age of female *Pseudoplusia includens* (Walker) to the daily oviposition rate. (After Mason and Mack, 1984.)

calculations. The peak oviposition rate for the mite species studied occurred at ca. 36° C, which coincided with the peak developmental rate observed for the immatures (Perring et al., 1984). A nonlinear model essentially identical in shape to equation 10.1 was employed by Mack et al. (1987) to describe the change in the proportion of the total number of *E. lignosellus* eggs laid per day with changing degree-days (fig. 10.6).

Food Quantity. The amount of food ingested affects many rate processes in insects. Some insects emerge as new adults with sufficient protein, lipid, and carbohydrate reserves to maturate their full compliment of eggs. For herbivorous species, an assumption of unlimited food supply is a sufficient approximation for most simulations. However, nutritional history of the immature stages may be important because fecundity of many insects is proportional to adult female weight (Southwood, 1978). The assumption of an unlimited food supply should therefore be approached with caution. The oviposition rate of herbivores at high densities, of most predators, and of many parasitic insects is limited by food quantity.

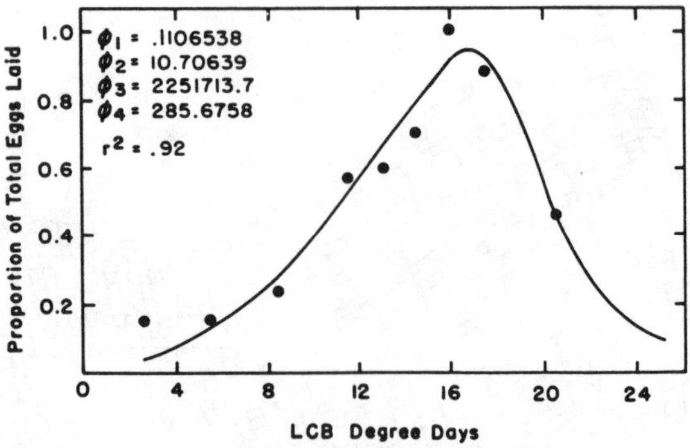

Figure 10.6–Nonlinear regression of the proportion of eggs laid versus degree-days for *Elasmopalpus lignosellus* (Zeller). A proportion of eggs of 1.0 equals the number of eggs per day laid at 30.5° C. (From Mack et al., 1987.)

Both linear and curvilinear relationships between food quantity and oviposition rate have been reported. Hassell (1978) assumed that the relationship between fecundity and food consumption could be described by:

$$F = y(I - b) \tag{10.2}$$

where

F = fecundity
y = a constant
I = ingestion rate
b = threshold ingestion rate below which reproduction ceases

A threshold ingestion rate was hypothesized because a proportion of food is necessary for maintenance respiration. Hassell (1978) stated that this model was well supported by several experiments, and cited the progeny production of *Podisus maculiventris* (Say) (Mukerji and Le Roux, 1969) as an example.

A more realistic model of the effects of food quantity on oviposition rate can be developed by assuming that with an unlimited amount of food and with other conditions being optimal, an insect will oviposit at a fixed maximum rate (fig. 10.7). Food will be limiting at less than this level, with oviposition decreasing at an increasingly faster rate as food becomes more and more limiting. Thus, the ovipositional curve should be shaped like a rectangular hyperbola:

$$\frac{d(O)}{dt} = \frac{(O_{max} X)}{(K_m + X)} \tag{10.3}$$

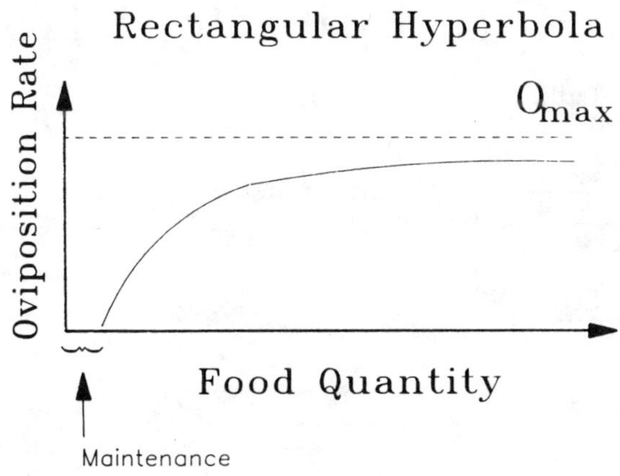

Figure 10.7–Relationship between food consumption and oviposition rate. O_{max} is the maximum oviposition rate.

where

$d(O)/dt$ = oviposition rate (no./time)
O_{max} = maximum oviposition rate
X = food quantity
K_m = a constant

This is a logical argument since growth equals feeding minus respiration, and in most adult insects all of the biomass for growth is allocated to progeny production. Feeding has been described for many insects by equation 10.3 (Holling, 1961), while respiration could be considered to be constant at a constant temperature. The result of equation 10.3 minus a constant (respiration) would be a rectangular hyperbola with a nonzero x-intercept, because of the amount of food needed for maintenance respiration. Rectangular hyperbolas are very common in nutrient-limited reactions, and have been used to describe relationships from prey-limited predator feeding rates (Hull et al., 1977) to photosynthesis in limited light conditions (Acock et al., 1983). All of these are enzyme-catalyzed reactions, and are best described by a hyperbolic function. A hyperbolic function for the description of the effect of food quantity on oviposition rate is supported by a wide range of examples (Hassell, 1978; Dover et al., 1979).

Food Quality. The quality of food is usually considered to be the nutritional value of the food source to an organism. Growth rate of many predaceous insects is known to vary with different prey species consumed (Hassell, 1978). For example, the growth rate of *Coleomegilla maculata lengi* Timberlake was higher when individuals were fed *Acyrthosiphon pisum* (Harr.) than on *Rhopalosiphum maidis* (Fitch) (Smith, 1965b). Assuming that oviposition is directly related to growth, oviposition rate in predaceous insects will vary with prey species. An example may be found in Smith (1965a), where

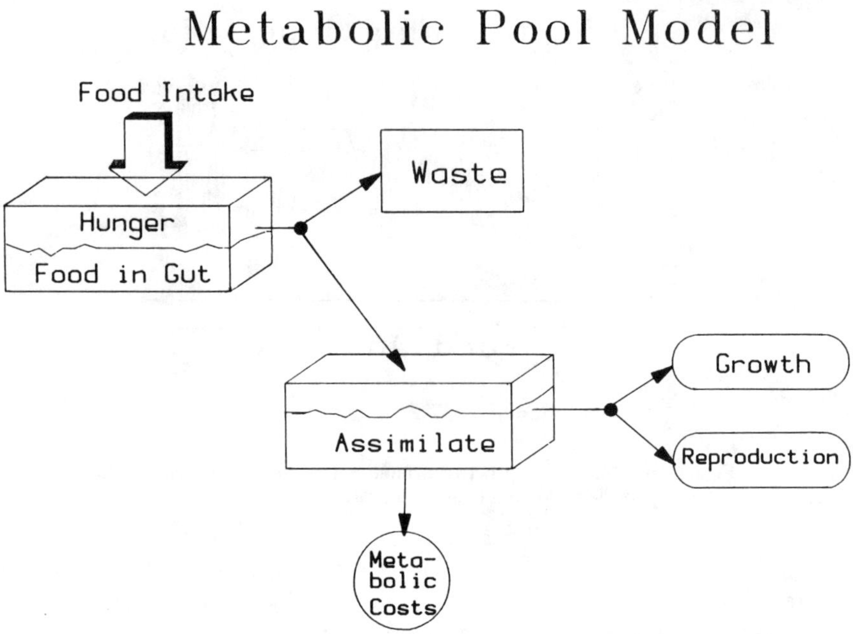

Figure 10.8–Metabolic pool model for a generalized predator. (Adapted from Gutierrez et al., 1981; used by permission of the authors.)

C. maculata lengi produced 5.7 eggs per milligram of food while feeding on *A. pisum*, 3.4 eggs/mg on *R. maidis*, and only 0.3 eggs/mg of food while feeding on *Aphis fabae* Scop.

The mathematics describing the switching of a predator from one species to another have not been completely elucidated (but, see Hassell, 1978, for proposed methods); consequently the effects of feeding on multiple prey species on oviposition rate have yet to be described. A possible approach to this problem can be found in Gutierrez et al. (1981), who developed a metabolic pool model for the growth, development, and reproduction of *Hippodamia convergens* Guérin-Méneville feeding on aphids (fig. 10.8). This model bases predator growth on biomass consumed, removes a proportion of the biomass consumed for metabolism, and allocates the rest to either growth or reproduction depending upon the individual's age. Another food source can easily be incorporated into this model by calculating the consumed biomass of the second food source.

Acceptability of Oviposition Sites. Many insects discriminate between optimal and suboptimal oviposition sites. Several species of parasitoids determine if a host has been previously handled by a conspecific (Vinson, 1976). For example, *Trichogramma evanescens* Westwood will not oviposit in host eggs visited by another *T. evanescens* (Salt, 1937). Equations for this type of interaction must take into account this discriminatory ability. Both Royama

(1971) and Rogers (1972) developed equations that allow for reduced parasite handling time with previously parasitized hosts. The Rogers (1972) equation, which has been the most commonly employed equation, is:

$$N_a = N_t \left\{ \frac{1 - \exp(-a\,t\,P_t)}{(1 + a t_h N_t)} \right\} \qquad (10.4)$$

where

N_a = number attacked
N_t = number of hosts at time "t"
a = search rate
t_h = handling time
P_t = number of parasitoids at time "t"

Curves generated from equation 10.4 are shaped similarly to those derived from equation 10.3. Equation 10.4 could be employed in a model to describe the effects of host density on oviposition rate of a parasitoid. In practice, both equations 10.3 and 10.4 must be fitted via nonlinear regression because linear transformations of these equations change the distribution of the data and, hence, the parameter values.

Evidence is accumulating that some herbivorous insects discriminate against oviposition sites containing one or more eggs. Oviposition deterrents have been found in weevils infesting buds or seeds (Giga and Smith, 1985), leaf-miners (Quiring and McNeil, 1984), tephritid flies (Prokopy et al., 1984), stored product pests (Corbet, 1973), and some lepidopterans (Dittrick et al., 1983). Few published models address this issue. McKibben et al. (1982) developed a boll weevil, *Anthonomus grandis grandis* Boheman, oviposition behavior model (fig. 10.9) that assumes the ability of female weevils to discriminate against infested cotton squares is a function of both infestation level and the number of discriminations made before laying an egg in an infested square. Model simulations indicate that discriminatory ability increases exponentially with increasing percent infestation, and the slope of the exponential curve increases with number of discriminations (fig. 10.10). Other models (e.g., Jones et al., 1975) assume there is a reduced probability that an egg will produce an adult insect if more than one egg is deposited in the same ovipositional site.

The suitability of various phenological ages of the ovipositional substrate is another factor that should be considered in modeling oviposition. On large plants, some herbivorous species change their oviposition sites as the plants mature. For example, three-cornered alfalfa hopper, *Spissistilus festinus* (Say), adults oviposit most of their eggs within lowermost nodes of young soybean plants, and as plants develop into more advanced vegetative stages, fewer eggs are deposited in the lower nodes (Rice and Drees, 1985). An example of a model containing a site selection component is the boll weevil model by Cate et al. (1979). Their approach is to assume that different sizes of cotton squares

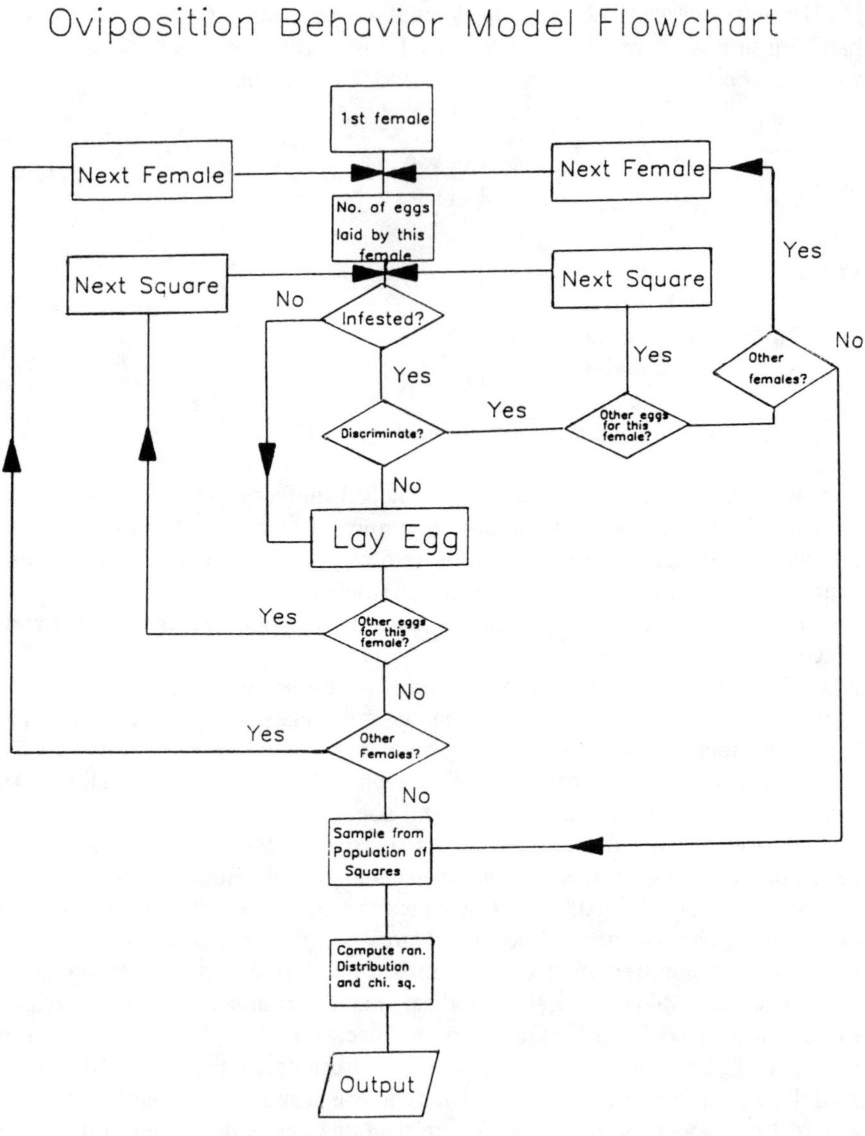

Figure 10.9–Flowchart of a model used to study oviposition behavior of boll weevils. (Adapted from McKibben et al., 1982; used by permission of the authors.)

have different acceptance probabilities by the female weevils. Given these probabilities, the availability of squares within a field determines the resultant egg distribution. The probability of no eggs laid is:

$$R = \sum (1 - p_i) f_i \tag{10.5}$$

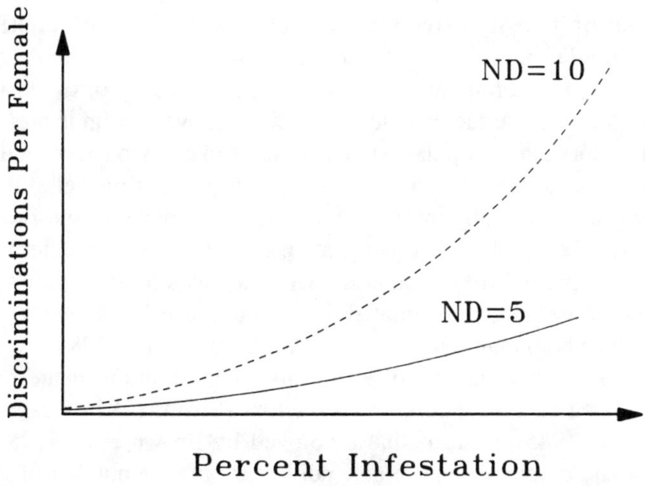

Figure 10.10–Discriminations among oviposition sites made by female boll weevils as a function of infestation level and number of discriminations before laying an egg in an infested square (ND). (Adapted from McKibben et al., 1982; used by permission of the authors.)

where

R = the probability that no egg is deposited on any given encounter with a square of any class

p_i = the probability that an ovipositing female accepts a fruit of size class "i" during a single encounter

f_i = field proportion of fruit of size class "i"

The probability of the egg being laid in a square of class "i", which shall be denoted as m_i, is:

$$m_i = \frac{f_i p_i}{\Sigma f_k p_k} \quad \text{for all "i"} \tag{10.6}$$

Note that there are multiple sets of p_i that will yield the same egg distribution m_i.

Availability of Mates. Mate density is a special case of the effects of population density on oviposition. A classic review of the effects of population density on oviposition has been published by Watt (1960). Density-dependent oviposition rates are well known in the literature. For example, Coulson et al. (1976) reported that *Dendroctonus frontalis* gallery length decreased exponentially with an increase in adult density. Foltz et al. (1976) reported that the number of eggs laid by *D. frontalis* increased linearly with gallery length, so in effect the number of eggs decreased exponentially with an increase in adult density.

Population density effects should be interpreted with caution, since increasing population size may affect sex ratio, adult male and female age structure, food quality, food availability, and the suitability of oviposition sites. An examination of these factors and their effect on oviposition is preferable to a lumping of factors into a population density term in an oviposition equation.

Mate density can have a large effect on population behavior. This is particularly true of parasitic insects which are arrhenotokous and produce only male progeny when females are not mated. For example, sex determination of parasitoid offspring of *Ophion flavidus* Brülle appears to be accomplished by a haploid/diploid mechanism. Unmated females produce 100% males, and mated females produce both males and females (Rohlfs and Mack, 1985).

Age-class matrix models offer a convenient solution to the problem of sex-differentiated survival and/or reproduction (Spain, 1982). Leslie age class models (Leslie, 1945), such as that employed by Brown et al. (1982), can be used to simulate both male and female adults, or only the number of females. A detailed discussion of population projection matrices and modifications can be found in Pielou (1977).

Verification of Oviposition Models

All insect population models need to be validated before their application in pest management programs. Validation is really a misnomer, because no model is ever validated for all sets of conditions in which the model may be used. A model can be verified. Model verification typically involves: 1) examining the validity of the assumptions; and 2) comparing model output behavior to "reality". The importance of examining the validity of the assumptions in an oviposition model should not be overlooked, because *the assumptions that were made and algorithms that were chosen dictate model behavior*. Model output can be compared to "reality" by comparing simulated oviposition curves to laboratory and field values of daily oviposition. Laboratory oviposition rates have been calculated for many species of insects; however, direct field measurements of oviposition are more difficult and in some instances impossible to measure.

Various methods have been developed for the estimation of oviposition when it cannot easily be measured by direct means. Southwood (1978) and Shoemaker et al. (1986) provide excellent reviews of this topic. These methods indirectly estimate population size. Actual estimates from field samples, such as those taken to validate the velvetbean caterpillar oviposition model (Barfield and Gregory, 1985) are much preferred. Also, simulation models can duplicate a wide variety of population curves because of the large number of parameters involved in a typical model. Thus, a similar pattern of occurred and expected population behavior from one data set does not imply true verification of model behavior. A model is verified to be correct only when simulated and observed oviposition are similar for a number of data sets, and when model assumptions have been shown to be reasonable, with little effect on model behavior.

Summary

Oviposition is the most commonly used recruitment in simulation models. Accurate estimates of oviposition are critical to the simulation of multi-generation insect pest population dynamics. Many factors should be considered in the development of oviposition models. Factors commonly included in oviposition models are the effects of adult age, temperature, food quantity, food quality, the acceptability of oviposition sites, and availability of mates. Mathematical relationships have been developed to describe the effects of these factors on oviposition.

Acknowledgment. We thank J. C. Allen, J. L. Goodenough, C. S. Barfield, and A. A. Berryman for their reviews of earlier drafts of this manuscript.

References

Acock, B. A., V. R. Reddy, F. D. Whistler, D. N. Baker, J. M. McKinion, H. F. Hodges and K. J. Boote. 1983. The soybean crop simulator Glycim: Model documentation 1982. Series: Response of Vegetation to Carbon Dioxide, No. 2. Washington, DC: USDA-ARS.

Barfield, C. S. and B. M. Gregory. 1985. The pattern of invasion by immigrant insect pest species. In *The Movement and Dispersal of Agriculturally Important Biotic Agents*, 233-248. Baton Rouge, LA: Claitor's Publishing Division.

Brown, G. C., A. A. Berryman and T. P. Bogyo. 1978. Simulating codling moth population dynamics: Model development, validation, and sensitivity. *Environmental Entomology* 7:219-227.

Brown, M. W., E. A. Cameron and F. M. Williams. 1982. Population model for the gypsy moth (Lepidoptera: Lymantriidae) egg parasite, *Ooencyrtus kuvanae* (Hymenoptera: Encyrtidae). *Environmental Entomology* 11:1209-1304.

Cate, J. R., G. L. Curry and R. M. Feldman. 1979. A model for boll weevil site selection. *Environmental Entomology* 8:917-921.

Corbet, S. A. 1973. Oviposition pheromone in larval mandibular glands of *Ephestia kuehniella*. *Nature* 243:537-538.

Coulson, R. N., A. M. Mayyasi, F. P. Hain and W. C. Martin. 1976. Resource utilization by the southern pine beetle, *Dendroctonus frontalis* (Coleoptera: Scolytidae). *Canadian Entomology* 108:353-362.

Dittrick, L. E., R. L. Jones and H. C. Chiang. 1983. An oviposition deterrent for the european corn borer, *Ostrinia nubilalis* (Lepidoptera: Pyralidae), extracted from larval frass. *Journal of Insect Physiology* 29:119-121.

Dover, M. J., B. A. Croft, S. M. Welch and R. L. Tummala. 1979. Biological control of *Panonychus ulmi* (Acarina: Tetranychidae) by *Amblyseius fallacis* (Acarina: Phytoseiidae) on apple: A prey-predator model. *Enviromental. Entomology* 8:282-292.

Flynn, P. W., R. A. J. Taylor and A. A. Hower. 1986. Predictive model for the population dynamics of potato leafhopper, *Empoasca fabae* (Homoptera: Cicadellidae), on alfalfa. *Enviromental. Entomology* 15:898-904.

Foltz, J. L., A. M. Mayyasi, F. P. Hain, R. N. Coulson and W. C. Martin. 1976. Egg-gallery length relationship and within-tree analyses for the southern pine beetle, *Dendroctonus frontalis* (Coleoptera: Scolytidae). *Canadian Entomology* 108:341-352.

Giga, D. P. and R. H. Smith. 1985. Oviposition markers in *Callosobruchus maculatus* F. and *Callosobruchus rhodesianus* Pic. (Coleoptera: Bruchidae). *Agricultural Ecosystem Environment* 12:229-233.

Gold, H. J. 1975. *Mathematical Modeling of Biological Systems — An Introductory Guidebook.* New York: John Wiley & Sons, Inc.

Greever, J. and G. P. Georghiou. 1979. Computer simulation of control strategies for *Culex tarsalis* (Diptera: Culicidae). *Journal of Medical Entomology* 3:180-188.

Gutierrez, A. P., J. B. Christensen, C. M. Merrit, W. B. Lowe, C. G. Summers and W. Cothran. 1976. Alfalfa and the egyptian alfalfa weevil (Coleoptera: Curculionidae). *Canadian Entomology* 108:635-648.

Gutierrez, A. P., J. U. Baumgaertner and K. S. Hagan. 1981. A conceptual model for growth, development, and reproduction in the ladybird beetle, *Hippodamia convergens* (Coleoptera: Coccinellidae). *Canadian Entomology* 113:21-33.

Hardman, J. M. 1978. A logistic model simulating environmental changes associated with the growth of populations of rice weevils, *Sitophilus oryzae*, reared in small cells of wheat. *Journal of Applied Ecology* 15:65-87.

Hassell, M. P. 1978. *The Dynamics of Arthropod Predator-Prey Systems*. Monographs in Population Biology 13. Princeton, NJ: Princeton University Press.

Holling, C. S. 1961. Principles of insect predation. *Ann. Rev. Entomology* 6:163-182.

Hull, L. A., D. Asquith and P. D. Mowery. 1977. The functional response of *Stethorus punctum* to densities of the european red mite. *Environmental Entomology* 6:85-90.

Hughes, R. D. and P. Sands. 1979. Modeling bushfly populations. *Journal of Applied Ecology* 16:117-139.

Jones, J. W., H. D. Bowen, R. E. Stinner, J. R. Bradley, Jr., R. S. Sowell and J. S. Bacheler. 1975. Female boll weevil oviposition and feeding processes: A simulation model. *Environmental Entomology* 4:815-821.

Leslie, P. H. 1945. The use of matrices in certain population mathematics. *Biometrika* 33:183-212.

Logan, J. A., D. J. Wollkind, S. C. Hoyt and L. K. Tanigoshi. 1976. An analytical model for description of temperature dependent rate phenomena in arthropods. *Environmental Entomology* 5:1133-1140.

McKibben, G. H., W. L. McGovern and W. A. Dickerson. 1982. Boll weevil (Coleoptera: Curculionidae) oviposition behavior: A simulation analysis. *Journal of Econ. Entomology* 75:928-931.

Mack, T. P. and Z. Smilowitz. 1982. CMACSIM, A temperature-dependent predator-prey model simulating the impact of *Coleomegilla maculata* (DeGeer) on green peach aphids on potato plants. *Environmental Entomology* 11:1193-1201.

Mack, T. P., J. W. Smith, Jr., and R. B. Reed. 1987. Mathematical model of the population dynamics of the lesser cornstalk borer, *Elasmopalpus lignosellus* (Lepidoptera: Pyralidae). *Ecological Modeling* 39:269-286.

Mason, L. J. and T. P. Mack. 1984. Influence of temperature on oviposition and adult female longevity for the soybean looper, *Pseudoplusia includens* (Walker) (Lepidoptera: Noctuidae). *Environmental Entomology* 13:379-383.

Mukerji, M. K. and E. J. Le Roux. 1969. A quantitative study of food consumption and growth in *Podisus maculiventris* (Hemiptera: Pentatomidae). *Canadian Entomology* 101:387-403.

Perring, T. M., T. O. Holtzer, J. L. Toole, J. M. Norman and G. L. Myers. 1984. Influences of temperature and humidity on pre-adult development of the banks grass mite (Acari: Tetranychidae). Environmental Entomology 13:338-343.

Pielou, E. C. 1977. *Mathematical Ecology*. New York: John Wiley & Sons, Inc.

Prokopy, R. J., B. D. Roitberg and A. L. Averill. 1984. Resource partitioning. In *Chemical Ecology of Insects*, ed. W. J. Bell and R. T. Carde, 301-330. Sunderland, MA: Sinauer Assoc.

Quiring, D. T. and J. N. McNeil. 1984. Intraspecific larval competition reduces efficacy of oviposition-deterring pheromone in the alfalfa blotch leafminer, *Agromyza frontella* (Diptera: Agromyzidae). *Environmental Entomology* 13:675-678.

Regan, T. E., R. E. Stinner, R. L. Rabb and C. Tuttle. 1979. A simulation model depicting the production of overwintering hornworms on tobacco. *Environmental Entomology* 8:268-273.

Rice, M. E. and B. M. Drees. 1985. Oviposition and girdling habits of the three-cornered alfalfa hopper (Homoptera: Membracidae) on preblooming soybeans. *Journal of. Econ. Entomology* 78:829-834.

Rogers, D. J. 1972. Random search and insect population models. *Journal of Animal Ecology* 43:239-253.

Rohlfs, W. M. and T. P. Mack. 1985. Seasonal parasitism rates, host size, and adult emergence pattern of parasitoids of the fall armyworm, *Spodoptera frigiperda* (J. E. Smith), with emphasis on *Ophion flavidus* Brülle (Hymenoptera: Ichneumonidae). *Annals of the Entomological Society of America* 78:217-220.

Royama, T. 1971. A comparative study of models for predation and parasitism. *Research Popul. Ecology Supplement* 1:1-91.

Salt, G. 1937. The sense used by *Trichogramma* to distinguish between parasitized and unparasitized hosts. *Proceedings of the Royal Society of London Serv. B.* 122:57-75.

Seawright, J. A., D. A. Dame and D. E. Weidhaas. 1977. Field survival and ovipositional characteristics of *Aedes aegypti* and their relation to population dynamics and control. *Journal of Med. Entomology* 37:62-70.
Sharpe, P. J. H. and D. W. DeMichele. 1977. Reaction kinetics of poikilotherm development. *Journal of Theoretical Biology* 64:649-670.
Shoemaker, C. A., G. E. Smith and R. G. Helgesen. 1986. Estimation of recruitment rates and survival from field census data with application to poikilotherm populations. *Agricultural Systems* 22:1-21.
Smith, B. C. 1965a. Effects of food on the longevity, fecundity, and development of adult coccinellids (Coleoptera: Coccinellidae). *Canadian Entomology* 97:910-919.
———. 1965b. Differences in *Anatis mali* Auct. and *Coleomegilla maculata lengi* Timberlake to changes in the quality and quantity of the larval food (Coleoptera: Coccinellidae). *Canadian Entomology* 97:1159-1166.
Southwood, T. R. E. 1978. *Ecological Methods with Particular Reference to the Study of Insect Populations*. London, England: Chapman & Hall.
Spain, J. D. 1982. *BASIC Microcomputer Models in Biology*. London, England: Addison-Wesley.
Toole, J. L., J. M. Norman, T. O. Holtzer and T. M. Perring. 1984. Simulating banks grass mite (Acari: Tetranychidae) population dynamics as a subsystem of a crop canopy-microenvironment model. *Environmental Entomology* 13:329-387.
Vinson, S. B. 1976. Host selection by insect parasitoids. *Ann. Rev. Entomology* 21:109-133.
Watt, K. E. F. 1960. The effect of population density on fecundity in insects. *Canadian Entomology* 92:674-695.
Wilkerson, G. C., J. W. Mishoe and J. L. Stimac. 1986. Modeling velvetbean caterpillar (Lepidoptera: Noctuidae) populations in soybeans. *Environmental Entomology* 15:809-816.

11

A Model for Long-distance Moth Dispersal

John K. Westbrook, Alton N. Sparks, Wayne W. Wolf,

Sammy D. Pair, and Jim R. Raulston

Introduction

Dispersive moth flights of several agriculturally important insect pests often occur suddenly, coinciding with development of favorable habitats downwind. Sparks (1978) described the significant mobility of *Heliothis* spp. and other noctuids as particularly important to population control programs. Blair and Catling (1974) documented a southward migratory path of African armyworm moths in southern Africa subsequent to larval outbreaks. Young (1979) postulated that insecticide-resistant fall armyworm larvae in the southeastern United States were progeny of moths which had migrated from areas south of Florida and/or Mexico. The spatial distribution of migrant moths varies with respect to time, making control measures expensive, and often, ineffective.

Egg and larval densities resulting from moth migrations can impose a serious stress on new host crops. For example, female fall armyworm moths oviposit at least 1000 eggs each. If 100 female fall armyworm moths migrated to a one-hectare cornfield and oviposited, their egg masses would be capable of generating 100,000 larvae/ha or about 2.5 larvae per stalk of corn. Although a farmer may be unable to detect nocturnal immigrations of 100 moths/ha, the farmer will surely see the subsequent damage caused by 100,000 larvae/ha.

Scientists have empirically linked weather variables to the phenological development and migration of insect pests. The flight activity of filbertworm moths in Oregon was influenced by temperature, wind, and rainfall (Aliniazee, 1983). Waterhyacinth moth dispersed northward through the Florida peninsula at an average rate of 1 km/day (Center, 1984). Domino et al. (1983) determined that early-spring populations of black cutworm in Iowa and northern Missouri were comprised only of immigrants from overwintering areas in southern states. Armyworm outbreaks in India were associated (Moray et al., 1983) with initial heavy precipitation followed by a break in monsoon activity. Warm temperature and abundant rainfall following cool winter temperature were cited (Luginbill, 1928) as typical precursors of severe fall armyworm infestations. Luginbill (1928) noted that periods of cool weather apparently suppressed populations of beneficial insects much more than populations of the fall armyworm prior to subsequent periods of warm weather. Also, migrant moths may have been responsible for increased fall armyworm populations in areas where warm, moist conditions promoted luxuriant crop growth.

Wind speed has generally been noted as the principal component of the moth transport vector. Arthur and Bauer (1981) concluded that sunflower moths migrated northward 2000 km in a few days to infest fields in Saskatchewan. *Heliothis armigera* migration en masse to the British Isles was associated with southerly and southeasterly wind which enabled displacements of as much as 2000 km (Pedgley, 1985). Shapiro (1980) determined that return southward migration of *Vanessa cardui* in northern California was a characteristic seasonal phenomenon, although densities were typically much less than those of spring northward migrations. Light trap captures on oil platforms in the Gulf of Mexico demonstrated that moths could disperse 160 km from shore with the assistance of synoptic scale wind patterns (Sparks et al., 1986). Tucker et al. (1982) documented the northwestward spread of armyworm moths in East Africa which coincided with the prevailing wind. Moth flight speeds of approximately 3 m/s have been documented by flight mills in laboratories (Carpenter et al., 1981). Typical wind speeds range from 5 m/s to a maximum of 28 m/s (Bonner, 1968) in clear, non-turbulent circulations. Thus, insect flight speed represents about 40% and 10% of the total moth transport vector in light and strong winds, respectively.

Scientists have applied atmospheric trajectories to chart probable paths of migrating moths (Muller and Tucker, 1986; Westbrook and Sparks, 1986). Mikkola and Albrecht (1986) concluded that radioactive moths which were captured in southern Finland had originated flight near Chernobyl following a nuclear power plant accident. Neck (1984) determined that snout butterflies migrated northward based upon successive observations across southern Texas. Pair et al. (1987) concluded that northeasterly wind following two cold front passages may have enabled corn earworm and fall armyworm moths to migrate southward from the Texas High Plains to the lower Rio Grande Valley, Texas. Mark-recapture experiments revealed that African armyworm moths had migrated downwind more than 90 km in one night (Rose et al., 1985). Larsen and Pedgley (1985) reported 1000-1300 km transport of butterflies across the Arabian Sea in connection with a slowly-moving tropical storm. Although trajectory methods have documented the likely mean direction and displacement of moth movement, no information has been provided about the density and distribution of migrant moths.

Long-distance dispersal models have calculated the spread of passive biotic agents such as foot-and-mouth disease (Gloster, 1983) and larval stage bagworms (Cox and Potter, 1986) but not the spread of active fliers. Stinner et al. (1983) emphasized the need for flexible simulation models to efficiently calculate insect dispersal from relatively small habitat areas. Knowledge of insect migratory flight behavior and population dynamics is a necessary component in the development of long-distance insect dispersal models.

Concurrent *in situ* measurements of insect flight activity and meteorological data have been compared to derive functional relationships.

Callahan et al. (1972) captured corn earworm moths at various levels on a 319-m tower to derive a vertical distribution of airborne moths. The height, density, speed, and extent of spruce budworm moth flight were documented by airborne radar measurements and correlated with simultaneous meteorological data (Greenbank et al., 1980). Drake and Farrow (1985) identified moth flights which exceeded a few hundred kilometers as determined by radar, aerial netting, and meteorological measurements. Glick and Noble (1961) conducted extensive aerial sampling of pink bollworm moths and wind velocity to indicate that these moths may be transported north-northeasterly from central Texas. Greenbank et al. (1980) used airborne radar to document the convergence and displacement of large numbers of airborne spruce budworm moths in the presence of wind shift fronts. Hendricks (1983) measured the effects of barometric pressure on survival and flight behavior of tobacco budworm moths to determine the resultant atmospheric impacts upon high-altitude, long-distance migrants. Lingren and Wolf (1982) investigated the nocturnal flight activity of tobacco budworm moths by radar and meteorological measurements. The dispersal of *Spodoptera exempta* moths observed by radar and an infra-red optical detector was determined to be strongly influenced by synoptic and mesoscale wind systems (Riley et al., 1979). Schaefer (1979) documented flight characteristics of spruce budworm moths in synoptic and mesoscale atmospheric circulations using an airborne radar. Knowledge of these factors generally requires special equipment such as entomological radars, aerial nets, atmospheric sounding instrumentation, and weather balloons (Sparks et al., 1985). Mean relationships such as migrant proportion of population, flight initiation time, flight altitude, flight threshold temperature, flight orientation, and flight duration can be examined by these measurements. Some of these factors may be determined alternatively in the laboratory for use in a dispersal model.

Concurrent moth flight data and meteorological data have been regressed to determine functional relationships. Hogg and Gutierrez (1980) quantitatively regressed beet armyworm flight phenology with weather variables. McCann et al. (1983) compared heat unit calculations with the regional spread of migrating cotton bollworm moths. Richmond et al. (1983) improved the prediction of spring flight of Nantucket pine tip moth relative to Julian date averages by calculating heat unit accumulations. The distribution and behavior of night-flying insects within a convergent wind system were simulated by Symmons and Luard (1982). For instance, a scientist may be able to determine that insect flight activity is a direct (positive) function of air temperature between upper and lower thresholds. Moth flight mill studies, radar observations, and fat body analyses have been used to estimate moth flight duration.

Insect-weather relationships can be incorporated into an atmospheric dispersal model. Rudd and Gandour (1985) applied concepts of diffusion theory to model insect dispersal for short distances. Such relationships modify the dispersal model to accommodate specific moth flight characteristics. A scientist may continue to add or change moth flight functional relationships to account for new knowledge in insect migration research.

This chapter describes the development of a modified atmospheric dispersal model which can be applied to the study of long-distance moth dispersal. The model is applied to a case study of meteorological conditions and a hypothetical *Heliothis zea* moth distribution. The dispersal model calculates the timing, location, and density of long-distance moth flights. The model is described sufficiently for new users to execute model calculations with or without specific modifications.

Procedure

Air temperature and wind speed data fields for the moth dispersal model were prepared from National Weather Service (NWS) upper-air data collected over a multi-state region. Values of the meteorological variables for a selected altitude (representing significant moth flight) were interpolated from the twice-daily NWS vertical atmospheric soundings. Linear interpolation was used for the existing dispersal model:

$$\hat{A} = \frac{A_{i-1} + \left(A_1 - A_{i-1}\right)\left(\hat{z} - z_{i-1}\right)}{\left(z_1 - z_{i-1}\right)} \tag{11.1}$$

where

A = meteorological variable
$\hat{A}$ = vertically interpolated value of the meteorological variable
z = altitude
i and (i−1) = the nearest measured atmospheric levels above (below) the selected altitude (e.g., z = 500 m)

Wind speed and air temperature data were then spatially interpolated to evenly spaced horizontal grid points, suggested here by a distance-weighted mean (Selby, 1974):

$$\hat{A}_{xy} = \frac{\left(\sum_{i=1}^{n} w_i \hat{A}_{z_i}\right)}{\left(\sum_{i=1}^{n} w_i\right)} \tag{11.2}$$

where

w = station-to-grid-point distance
i = station number
n = total number of interpolating stations (e.g., three for the model used in this chapter)
xy = a specific horizontal grid point

Because wind velocity is a vector quantity (i.e., possesses speed and direction), a horizontal smoothing technique, such as successive relaxation (Holton, 1972), was necessary to reduce computational divergence/convergence.

Migratory moth field densities were hypothesized to initiate the dispersal model calculations. In this model, a hypothetical moth concentration maximum was extrapolated to grid points via an exponential decay function:

$$\hat{C}_{xy} = C \exp\left(-\left(\frac{w}{(2\Delta x)}\right)^2\right) \quad (11.3)$$

where C is the adult field density at a source (e.g., 1000 moths/ha density at Brownsville, Texas) and Δx is the horizontal grid spacing (e.g., ca. 64 km).

A uniform moth concentration equal to the county midpoint value was assigned to each county. Hypothetical moth concentrations were daily interpolated to county midpoints and added to the previous county population of immigrant moths. The hypothetical adult field density at the source (i.e., 1000/ha-day) represents measured field density of corn earworm in the lower Rio Grande valley (Raulston et al., 1986) rounded up to the nearest power of ten. The 64-km horizontal grid spacing was simply a carryover from an earlier (non-graphic) form of the dispersal model computer program.

A Gaussian atmospheric dispersion model was chosen as the framework for the long-distance insect dispersal model. The Gaussian adjective means that the model uses a fixed reference with which to calculate dispersed moth movements. The Gaussian reference is the antithesis of the Lagrangian reference which sets the reference to the center-of-mass of a dispersing medium. The Gaussian reference was chosen for this model development because the dispersing medium (i.e., moths) do not typically organize into a symmetrical region. The Lagrangian reference requires that the dispersing medium have a meaningful center-of-mass of determining the local dispersion.

The Gaussian atmospheric dispersion equation was comprised of a linear combination of input and output factors. Simply explained, the time change (increase or decrease) of moth density at a locality equaled the moth influx/outflux due to horizontal advection (transport), horizontal dispersion (spreading), sources of migrant moths, and sinks of migrant moths (e.g., washout by precipitation, predation, and flight termination). As the reader will note, sources and sinks will vary for different insect pests. Mathematically, the dispersion equation takes the following form:

$$\frac{\partial C}{\partial t} = \left(\frac{u\partial C}{\partial x} + \frac{v\partial C}{\partial y}\right) + \left[\frac{\partial}{\partial x}\left(K_x \frac{\partial C}{\partial x}\right) + \frac{\partial}{\partial y}\left(K_y \frac{\partial C}{\partial y}\right)\right] + S^+ - S^- \quad (11.4)$$

The left-side term is the local time rate-of-change $\partial(\,|\,)/\partial t$ of the moth density (C). The local time rate-of-change is thus equal to the contributions due to horizontal advection, horizontal dispersion, sources, and sinks which are the first, second, third, and fourth right-side terms, respectively. The partial derivatives ($\partial/\partial x$, $\partial/\partial y$, and $\partial/\partial t$) represent the ascendant of a quantity with respect to the single denominator variable only. The terms K_x and K_y represent the longitudinal and latitudinal eddy diffusivity coefficients, respectively. The insect source (S^+) may be described as a function of field densities, environmental conditions, and time of night. Insect sinks (S^-) may be linearly combined into this single term, if desired. Further information on the use of the Gaussian dispersion model is described by Westbrook (1985).

Results

A hypothetical case is presented here to clarify the use of this insect dispersal model. Based upon field observations of *Heliothis* spp. flight with radar in southern Texas and other observations in the laboratory, the following model assumptions were employed. Moths frequently congregated at altitudes of approximately 500 m above ground level (AGL), a height which approximates the midpoint of the planetary boundary layer (PBL). Thus, the dispersal model assumed a flight altitude of 500 m AGL (Wolf et al., 1986) and meteorological data were interpolated to this height. Moth flight was assumed to continue for a maximum of five hours based upon flight mill studies (Carpenter et al., 1981) and entomological radar observations of moth flight activity (Wolf et al., 1986). Moth flight was terminated for air temperature at or below 12° C (Carpenter et al., 1981; Hartstack et al., 1982; Wolf et al., 1986). This model assumed no vertical convection or vertical diffusion in the nocturnal PBL. A fixed horizontal eddy diffusivity of 100 m^2/s was assumed similar to PBL measurements by Lanham (1980). The termination of moth flight due to flight duration and threshold air temperature were the sole moth sinks considered here. Repeated nocturnal flights by a 50% migrant proportion of the total moth population on successive nights were assumed.

For the hypothetical moth dispersal, a source of 1000 migrant moths/ha-day was assumed at Brownsville, Texas. This migrant moth density was extrapolated to surrounding counties to depict a regional migrant moth population (fig. 11.1). Note the initial three-orders-of-magnitude insect concentration decrease within a 320-km distance from the hypothetical source center (Brownsville, Texas) based upon calculations using the forementioned exponential decay function for $\hat{C}_{xy}$.

Migrant moth dispersal was calculated for a five-hour period or until the air temperature decreased to 12° C, whichever occurred first. Airborne moths at a specific grid point were assumed to instantaneously descend to the surface when either of these two criteria existed. The moth dispersal criteria are flexible within the present computer model (e.g., the program user may select a maximum flight duration of eight hours and a minimum flight temperature of 6° C).

Moth dispersal maps (fig. 11.2) were prepared to document the direction, distribution, and density of dispersed moth movements on a county basis.

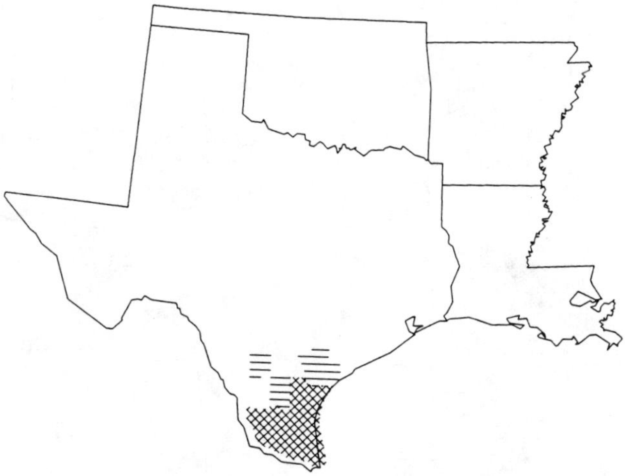

Figure 11.1–Simulated initial concentration of *Heliothis zea* moths valid at 1900 CDT on 19 May 1984. (Blank, line, cross-hatched, and solid areas represent 0-10, 10-100, 100-1000, and 1000-10,000 moths/ha.)

Successive dispersal maps revealed a chronological pattern of regional migrant moth populations. The initial dispersal map (20 May) revealed a northward spread of approximately 250 km. Strong, southerly wind in advance of an eastward moving cold front (fig. 11.3a) corroborated the simulated northward moth displacement on the night of 19 May 1984. The cold front moved into south-central Texas on 20 May 1984 (fig. 11.3b), and westerly wind displaced the western perimeter of the simulated moth population eastward (21 May). The cold front stalled over south-central Texas on 21 May 1984 (fig. 11.3c) to direct southerly wind across Louisiana, Arkansas, and the coastal area of Texas and displaced the simulated moth population northwestward along the Rio Grande valley and northward into eastern Texas (22 May). A second cold front stalled across the Texas panhandle on 22 May 1984 (fig. 11.3d) and extended the simulated moth distribution approximately 250 km due north (23 May). The front moved southward into central Texas on 23 May 1984 (fig. 11.3e) and appeared to redistribute the simulated moth population within the same geographic area (24 May) as on 22 May 1984. Southerly wind re-established in Oklahoma and northern Texas on 24 May 1984 (fig. 11.3f) and displaced the simulated moth distribution northward into central Oklahoma and eastward into southwest Louisiana (25 May).

Discussion

This chapter has presented a technique which is capable of quantifying the direction, distribution, and density of long-distance moth dispersal. The simulated dispersal maps showed a conceptually reasonable correlation with

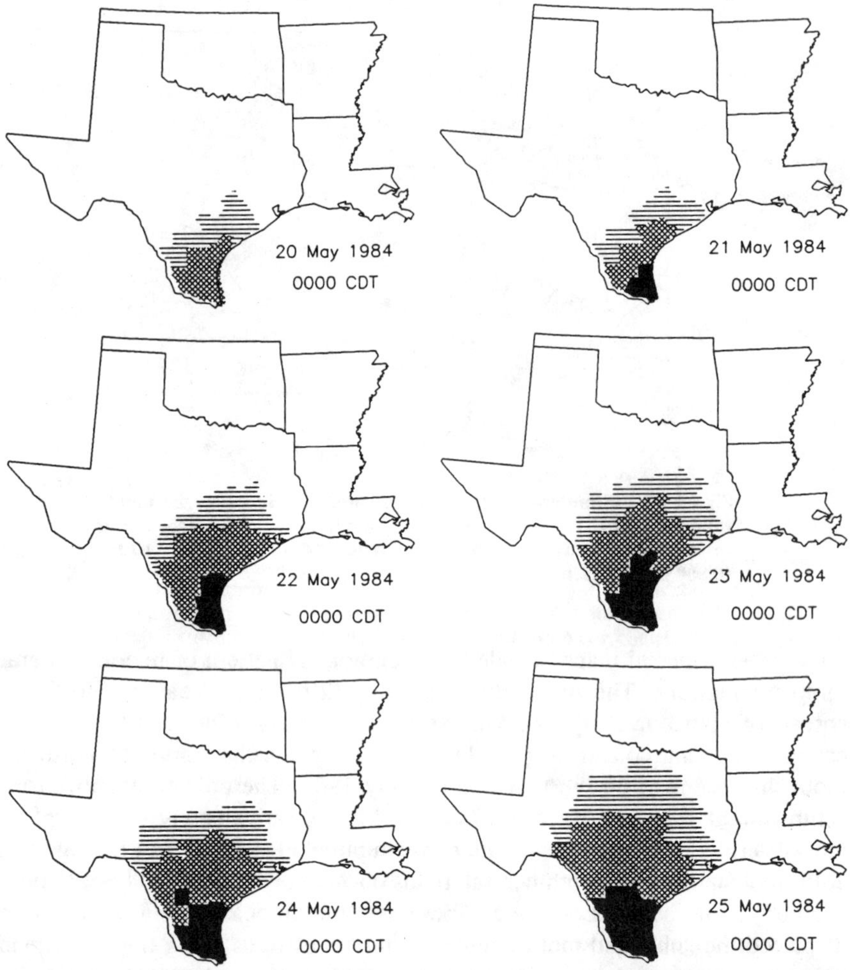

Figure 11.2–Simulated successive, five-hour nocturnal atmospheric dispersion of *Heliothis zea* moths at 500 m above ground level from the Lower Rio Grande valley, Texas. (Blank, line, cross-hatched, and solid areas represent 0-10, 10-100, 100-1000 and 1000-10,000 moths/ha.) A proportion of 50% migrant moths, a threshold air temperature of 12° C, and a daily source of one thousand moths per hectare at Brownsville, Texas (spatially extrapolated via an exponential decay function), were assumed for the simulated dispersion.

synoptic weather events. Moth dispersal simulation modeling may benefit the economic planning of effective insect migration research activities and integrated pest management strategies.

More extensive laboratory and field observations of moth flight behavior should be collected and interpreted. These observations will increase the

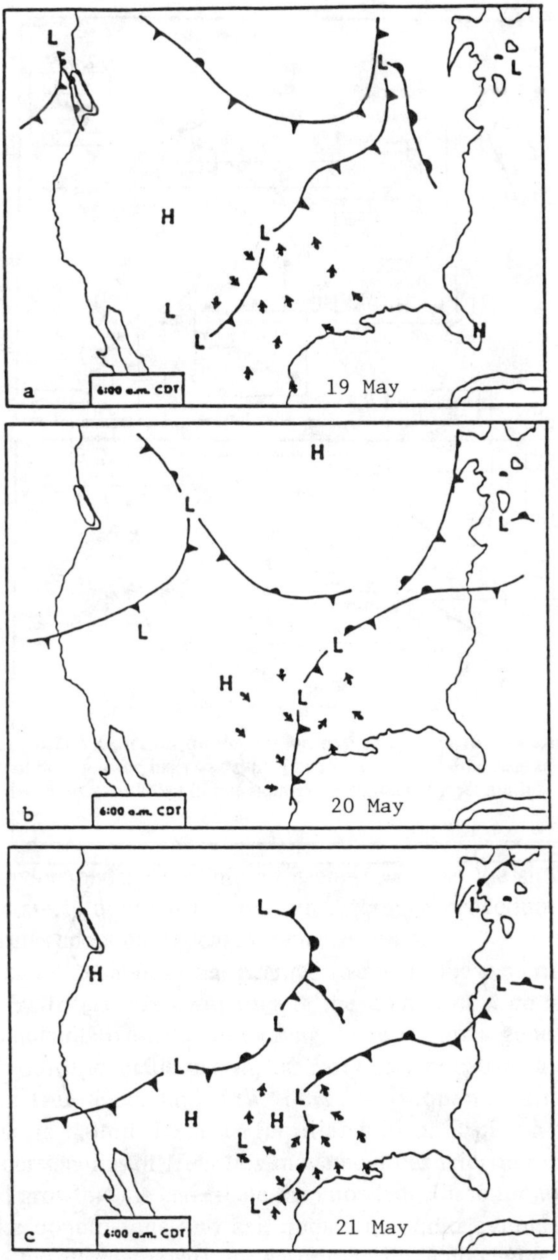

Figure 11.3–Daily surface synoptic weather maps for the continental U.S. for 19-24 May 1984 (a-f, respectively). Arrows represent the prevailing wind direction in the south-central United States (Continued on page 180).

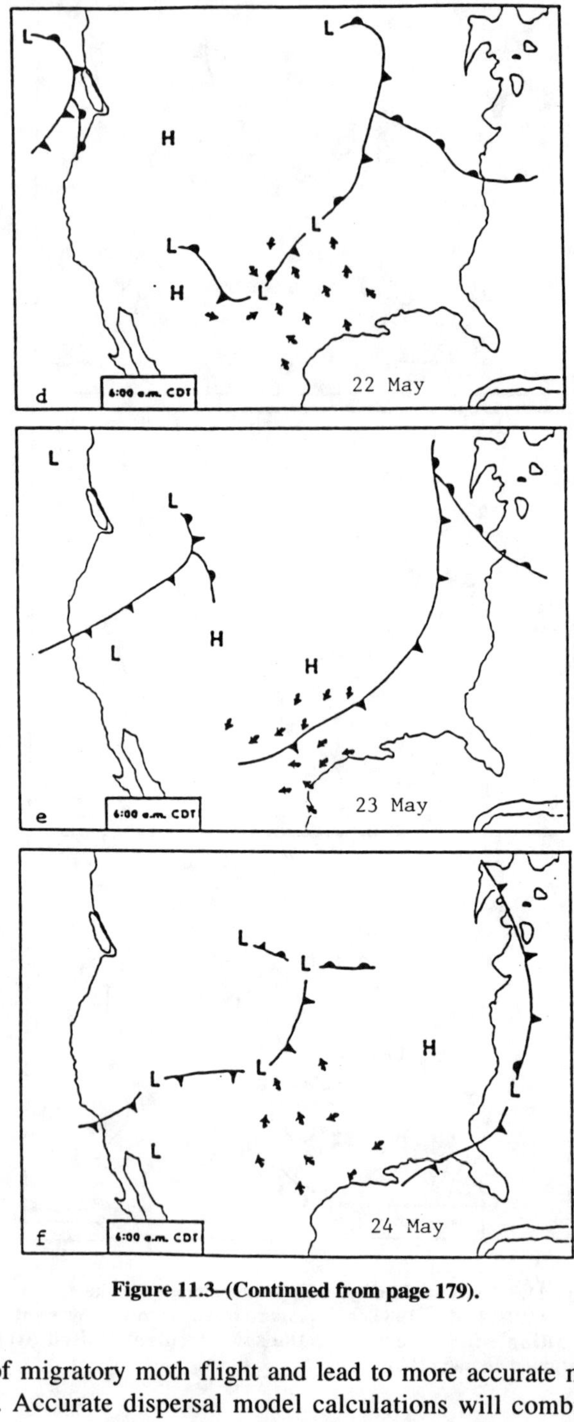

Figure 11.3–(Continued from page 179).

knowledge of migratory moth flight and lead to more accurate moth dispersal calculations. Accurate dispersal model calculations will combine well with existing insect phenology models (Hartstack et al., 1983) and crop growth models (Andrews et al., 1984). An integrated model which simulates host crop development, insect phenology, and migratory moth flight may significantly improve existing pest management strategies that solely evaluate indigenous pest populations.

Extensive meteorological measurements should be made in the nocturnal PBL to disclose the mean transport wind velocity and turbulence which impacts noctuid moth dispersal. An array of doppler radars may soon be operational to measure wind profiles in the central United States at 0.5-hour intervals rather than 12-hour intervals (Kaimal, 1987). Atmospheric trajectories could be estimated by tracking a single superpressure balloon via radar or LORAN-C (Clawson et al., 1986). Radar or LORAN-C tracking of multiple superpressure balloons could yield an estimate of atmospheric dispersion. These measurements will significantly improve the atmospheric component of moth dispersal estimates.

References

Aliniazee, M. T. 1983. Influence of environmental factors on daily flight activity of the filbertworm, *Melissopus latiferreanus* (Lepidoptera, Olethreutidae). *Ecological Entomology* 8:241-248.

Andrews, G., L. G. Brown and K. S. Akbay. 1984. Computer assisted cotton insect management. In *Proceedings of the Beltwide Cotton Production Research Conferences*, pp. 239-244. Memphis, TN: National Cotton Council.

Arthur, A. P. and D. J. Bauer. 1981. Evidence of the northerly dispersal of the sunflower moth by warm winds. *Environmental Entomology* 10:528-533.

Blair, B. W. and H. D. Catling. 1974. Outbreaks of African armyworms, *Spodoptera exempta* (Walker) (Lepidoptera, Noctuidae), in Rhodesia, South Africa, Botswana and Southwest Africa from February to April 1972. *Rhodesian Journal of Agricultural Research* 12:57-67.

Bonner, W. D. 1968. Climatology of the low level jet. *Monthly Weather Review* 96:833-850.

Callahan, P. S., A. N. Sparks, J. W. Snow and W. W. Copeland. 1972. Corn earworm moth: Vertical distribution in nocturnal flight. *Environmental Entomology* 1:497-503.

Carpenter, J. E., A. N. Sparks and E. A. Harrell. 1981. Effects of temperature on wingbeat frequency and sustained flight on certain lepidopteran insects. *Journal of Georgia Entomology Society* 16:451-457.

Center, T. D. 1984. Dispersal and variation in infestation intensities of waterhyacinth moth, *Sameodes albiguttalis* (Lepidoptera: Pyralidae), populations in peninsular Florida. *Environmental Entomology* 13:482-491.

Clawson, K. L., C. R. Dickson, E. F. Pound and G. E. Start. 1986. Long-range tetroon tracking with LORAN-C transponders. In Extended Abstracts, American Meteorol. Society, 5th Joint Conference on Air Pollution Meteorology, 363-365. Chapel Hill, NC.

Cox, D. L. and D. A. Potter. 1986. Aerial dispersal behavior of larval bagworms, *Thyridopteryx ephemeraeformis* (Lepidoptera: Psychidae). *Canadian Entomology* 118:525-536.

Domino, R. P., W. B. Showers, S. E. Taylor and R. H. Shaw. 1983. Spring weather pattern associated with suspected black cutworm moth introduction to Iowa. *Environmental Entomology* 12:1863-1872.

Drake, V. A. and R. A. Farrow. 1985. A radar and aerial-trapping study of an early spring migration of moths (Lepidoptera) in inland New South Wales. *Australian Journal of Ecology* 10:223-235.

Glick, P. A. and L. W. Noble. 1961. Airborne movement of the pink bollworm and other arthropods. USDA Technical Bulletin No. 1255.

Gloster, J. 1983. Forecasting the airborne spread of foot-and-mouth disease and Newcastle disease. *Philosoph. Trans. Royal Society of London, Ser. B.* 302:535-541.

Greenbank, D. O., G. W. Schaefer and R. C. Rainey. 1980. Spruce budworm (Lepidoptera: Tortricidae) moth flight and dispersal: New understanding from canopy observations, radar, and aircraft. *Mem. Entomology Society of Canada* 110:1-49.

Hartstack, A. W., J. D. Lopez, R. A. Muller, W. L. Sterling, E. G. King, J. A. Witz and A. C. Eversull. 1982. Evidence of long range migration of *Heliothis zea* (Boddie) into Texas and Arkansas. *Southwestern Entomology* 7:188-201.

Hartstack, A. W., J. D. Lopez and J. A. Witz. 1983. Potential use of models to study insect migration. ASAE Paper No. 83-1502. St. Joseph, MI: ASAE.

Hendricks, D. E. 1983. Effects of high-altitude exposure on survival and behavior of tobacco budworms (Lepidoptera: Noctuidae) and boll weevils (Coleoptera: Curculionidae). *Environmental Entomology* 12:1897-1902.

Hogg, D. B. and A. P. Gutierrez. 1980. A model of the flight phenology of the beet armyworm (Lepidoptera: Noctuidae) in central California. *Hilgardia* 48:1-36.

Holton, J. R. 1972. *An Introduction to Dynamic Meteorology*. New York: Academic Press.

Kaimal, L. 1987. Some profiler history. *NOAA/ERL Profiler Forum* (March):1-2.

Lanham, L. 1980. Wintertime dispersion processes in the Lake Tahoe basin. In *Proceedings of the American Meteorol. Society, 2nd Conf. on Applications of Air Pollution Meteorology*, 520-525. New Orleans, LA.

Larsen, T. B. and D. E. Pedgley. 1985. Indian migrant butterflies displaced to Arabia by monsoon storm 'Aurora' in August 1983. *Ecological Entomology* 10:235-238.

Lingren, P. D. and W. W. Wolf. 1982. Nocturnal activity of the tobacco budworm and other insects. In *Biometeorology in Integrated Pest Management*, eds. J. L. Hatfield and I. J. Thomason, 211-228. San Diego, CA: Academic Press, Inc.

Luginbill, P. 1928. The fall armyworm. USDA Technical Bulletin No. 34.

McCann, I. P., M. J. McFarland and A. W. Hartstack. 1983. Meteorological parameters for modeling migration of the cotton bollworm. ASAE Paper No. 83-1001. St. Joseph, MI: ASAE.

Mikkola, K. and A. Albrecht. 1986. Radioactivity in Finnish night-flying moths (Lepidoptera) after the Chernobyl accident. *Notulae Entomology* 66:153-157.

Moray, P. E., S. K. Kazi and A. S. Ballal. 1983. Influence of weather factors on the outbreak of armyworms. *Journal of Maharashtra Agricultural University* 8:128-131.

Muller, R. A. and N. L. Tucker. 1986. Climatic opportunities for the long-range migration of moths. In *Long-Range Migration of Moths of Agronomic Importance to the United States and Canada: Specific Examples of Occurrence and Synoptic Weather Patterns Conducive to Migration*, ed. A. N. Sparks, 61-83. USDA-ARS, ARS-43.

Neck, R. W. 1984. On the origin of snout butterflies (*Libytheana bachmanii larvata*, Libytheidae) in a 1978 migration in southern Texas. *Journal of Lepidopterists' Society* 38:319-322.

Pair, S. D., J. R. Raulston, D. R. Rummel, J. K. Westbrook, W. W. Wolf, A. N. Sparks and M. F. Schuster. 1987. Development and production of corn earworm and fall armyworm in the Texas High Plains: Evidence for reverse fall migration. *Southwestern Entomology* 12:89-99.

Pedgley, D. E. 1985. Windborne migration of *Heliothis armigera* (Hubner) (Lepidoptera: Noctuidae) to the British Isles. *Entomology Gazette* 36:15-20.

Raulston, J. R., S. D. Pair, F. A. Pedraza Martinez, J. Westbrook, A. N. Sparks and V. M. Sanchez Valdez. 1986. Ecological studies indicating migration of *Heliothis zea, Spodoptera frugiperda*, and *Heliothis virescens* from northeastern Mexico and Texas. In *Insect Flight: Dispersal and Migration*, ed. W. Danthanarayana, 204-220. Berlin, West Germany: Springer-Verlag Publishing.

Richmond, J. A., H. A. Thomas and H. Bhattacharyya. 1983. Predicting spring flight of Nantucket pine tip moth (Lepidoptera: Olethreutidae) by heat unit accumulation. *Journal of Econ. Entomology* 76:269-271.

Riley, J. R., D. R. Reynolds and M. J. Farmery. 1979. Radar observations of *Spodoptera exempta* Kenya, March-April 1979. Centre for Overseas Pest Research Report 54:1-43.

Rose, D. J. W., W. W. Page, C. F. Dewhurst, J. R. Riley, D. R. Reynolds, D. E. Pedgley and M. R. Tucker. 1985. Downwind migration of the African armyworm moth, *Spodoptera exempta*, studied by mark-and-recapture and by radar. *Ecological Entomology* 10:299-313.

Rudd, W. G. and R. W. Gandour. 1985. Diffusion model for insect dispersal. *Ecological Entomology* 78:295-301.

Schaefer, G. W. 1979. An airborne radar technique for the investigation and control of migrating pest insects. *Phil. Transactions of the Royal Society of London* 287:459-465.

Selby, S. M., ed. 1974. *Standard Mathematical Tables,* 22nd Ed. Cleveland, OH: CRC Press.

Shapiro, A. M. 1980. Evidence for a return migration of *Vanessa cardui* in northern California (Lepidoptera: Nymphalidae). *Pan-Pacific Entomology* 56:319-322.

Sparks, A. N. 1978. Migration of *Heliothis* spp. and other noctuids (in relation to crop pest control). *Folia Mexicana* 131-132.

Sparks, A. N., W. W. Wolf, J. K. Westbrook, J. R. Raulston and S. D. Pair. 1985. Ideal components of a project investigating the movement and dispersal of noctuids. In *The Movement and Dispersal of Agriculturally Important Biotic Agents*, eds. D. R. MacKenzie, C. S. Barfield, G. G. Kennedy and R. D. Berger, 501-507. Baton Rouge, LA: Claitor's Publishing Division.

Sparks, A. N., R. D. Jackson, J. E. Carpenter and R. A. Muller. 1986. Insects captured in light traps in the Gulf of Mexico. *Annals of Entomological Society of America* 79:132-139.

Stinner, R. E., C. S. Barfield, J. L. Stimac and L. Dohse. 1983. Dispersal and movement of insect pests. *Ann. Rev. Entomology* 28:319-335.

Symmons, P. M. and E. J. Luard. 1982. The simulated distribution of night-flying insects in a wind convergence. *Australian Journal of Zoology* 30:187-198.

Tucker, M. R., S. Mwandoto and D. E. Pedgley. 1982. Further evidence for windborne movement of armyworm moths, *Spodoptera exempta*, in East Africa. *Ecological Entomology* 7:463-473.

Westbrook, J. K. 1985. Measurements and techniques to document aerobiological transport systems. In *The Movement and Dispersal of Agriculturally Important Biotic Agents*, eds. D. R. MacKenzie, C. S. Barfield, G. G. Kennedy and R. D. Berger, 219-229. Baton Rouge, LA: Claitor's Publishing Division.

Westbrook, J. K. and A. N. Sparks. 1986. The role of atmospheric transport in the economic fall armyworm (Lepidoptera: Noctuidae) infestations in the southeastern United States in 1977. *Florida Entomology* 69:492-502.

Wolf, W. W., J. K. Westbrook and A. N. Sparks. 1986. Relationship between radar entomological measurements and atmospheric structure in south Texas during March and April 1982. In *Long-Range Migration of Moths of Agronomic Importance to the United States and Canada: Specific Examples of Occurrence and Synoptic Weather Patterns Conducive to Migration*, ed. A. N. Sparks, 84-97. USDA-ARS, ARS-43.

Young, J. R. 1979. Assessing the movements of the fall armyworm (*Spodoptera frugiperda*) using insecticide resistance and wind patterns. In *Movement of Highly Mobile Insects: Concepts and Methodology in Research*, eds. R. L. Rabb and G. G. Kennedy, 344-351. North Carolina State University, Raleigh: University Graphics.

12

Integrated Decision Model for Velvetbean Caterpillar Control

J. Wayne Mishoe and Gail G. Wilkerson

Introduction

There is a need to improve the way farm level crop management decisions are being made. The opportunity to change has been stimulated by the increasing knowledge of crop system behavior, and by external pressures brought to bear on the farmer by low profit margins, by environmental regulations, and by the difficulty in simultaneously considering these complex issues in making rational decisions. State and Federal agencies need the capability to analyze the impact of governmental regulations on agricultural productivity, resource use, and natural environments. In each of these cases, estimates of crop yield in response to weather and management practices are required. To do this in the most effective way requires the use of crop growth models which integrate the effects of various environmental and management inputs and predict the resulted yield. The example scenario presented in this study focuses on the implementation of a soybean growth model, weather information, and several component models linked with the soybean model (Mishoe et al., 1984).

The soybean growth model of choice for this study is SOYGRO (Wilkerson et al., 1983a). There are several other crop models available for soybeans and other crops, however, only a few have been developed to the extent that they can be used directly by growers and other users (McClendon et al., 1979; Arkin et al., 1980; Arkin and Dugas, 1981; Meyer and Curry, 1981). SOYGRO has been implemented in a research model called SICM and includes component models of Velvetbean Caterpillar (VBC) (Wilkerson et al., 1986) the southern green stinkbug (SGSB) (Rudd, 1975), the corn earworm (CEW) (Stinner et al., 1975) and a soil water model. A second implementation called PESTDEC (Wilkerson and Mishoe, 1988) focuses on real-time decision making.

The types of decisions that a farm manager must make can be classified into two broad classes of within season and preseason, assuming the season begins at planting. The types of information needed to make each of these classes of decisions vary. For example, one preseason decision that a producer must make is the number of hectares of each crop that he will plant. For this decision he must know the expected production cost, selling price, and expected yield for each crop for each management scenario. The ability to estimate these values accurately affects the value of a management decision. Weather variability, however, causes the greatest variation in yield between years. Thus the ability to integrate weather information into the decision process can greatly improve one's ability to forecast yield. The approach presented in this study defines one implementation of the soybean model that can be used to study the effects of weather on crop management decisions and to provide the producer with better information upon which to make decisions. However, any such implementation will continue to require assessment by the producer and can only serve as a tool providing information presented in more specific detail than can be acquired through other means.

The implementation of PESTDEC is designed to help answer the question, "Should I apply pesticide today?". The process models simulate the effects of weather and management practices on the pest populations and on crop growth, and present the user with estimates of expected profit and associated risk. In many respects the problem is similar to the irrigation model problem presented by Swaney et al. (1983). Their model incorporated the uncertainty of yield and profit caused by uncertain future weather conditions. With insect pest there is additional uncertainty about future pest populations which cannot be linked directly to weather. The variability of pest flux into and out of a particular field and the variability of pest mortality may in fact be correlated with weather factors and crop production. However, in most cases this information has not been developed. For this reason, PESTDEC incorporated the uncertainty about future pest populations into its decision algorithm.

Making a pest management decision is further complicated by the fact that the question, "Should I spray today?", is not the only question that must be answered. The related questions, "If I spray today, what pesticides should I use and at what rate should I apply it?", must be answered. Since the costs of pesticide vary and different pesticides have quite different effects on pest populations and on the natural enemies of the pest populations, the selection of a particular pesticide can affect the answer to the question whether it is better to spray or to wait and evaluate the decision again at a later date.

Decision Model Description

PESTDEC (Wilkerson and Mishoe, 1988; Wilkerson et al., 1983b) is an interactive decision model which uses the soybean crop growth model SOYGRO (Wilkerson et al., 1983a), soil water balance and economic subroutines described in part by Swaney et al. (1983), and a velvetbean caterpillar (VBC) population dynamics model (Wilkerson et al., 1986). The larval stages feed on soybean leaves and may completely defoliate a field in a matter of days if the population level is high and the field is not treated. At present VBC is the only pest included in the model, but the structure is such that other pests can be added as appropriate population models become available.

Simulation from Planting to Decision Date. Figure 12.1 is a schematic of the basic steps of the decision-making algorithm. The user of PESTDEC must first specify today's date, the date upon which a decision is to be made, and specify the field for which the decision is to be made. A report on information stored in data files for that field is displayed, and the user is given the opportunity to change or add to that information. For instance, he may enter the dates and amounts of any irrigations that have been made to the field between the last time he ran the model for this field and today.

When the field information has been updated, weather data for the season prior to today are read from a data file and crop growth is simulated from planting until today. A crop and pest status report is displayed and the user is given the option of looking at graphs of a number of crop and pest state variables. The assumption is that the model will be used in conjunction with pest scouting reports. If there are scouting data available for the field being simulated, they are displayed along with the simulated values of the state variable being graphed (fig. 12.2).

After the user has looked at graphs of simulated and sampled pest populations, he may decide that the simulated population does not reflect the true state of the population and he may change the adult VBC influx parameters. There is no way to actually measure influx into the field at present, and varying influx parameters is the simplest way to match simulated to observed larval populations. The user may change the amplitude of the adult influx, the starting date, or the length of time over which the influx occurs. If this is the first time that a particular field has been simulated, an adult influx has been assumed which yields the average number of larvae, with the larval population peak occurring at the normal time of year. After any changes in influx parameters, the model is reinitialized and the simulation from planting until today may be repeated until a satisfactory fit between simulated and observed populations is obtained (fig. 12.3).

The user is then given the opportunity of changing the values of a number of important state variables on today's date. For instance, if he has an estimate of leaf area index (LAI) or percent defoliation today, and the simulated value for today is different, he can change the simulated value in a manner similar to that described by Arkin et al. (1980).

The goal of the above procedures is to match the simulation to the particular field situation as closely as possible on today's date. The crop model is used to integrate this information about the past to obtain an accurate picture of the state of the crop today before advancing into the uncertain future.

Algorithm for Optimization. The algorithm developed in this decision model is similar to the open loop stochastic control problem as described by Zavaleta et al. (1980). Here we will start with an open loop control problem statement for deterministic and stochastic cases, then show how the PESTDEC algorithm relates to this class of problems.

For the deterministic case, we first assume that we have perfect knowledge at time to (today's date) about weather and pest population influxes for the rest of the season. We also assume that economic factors, such as

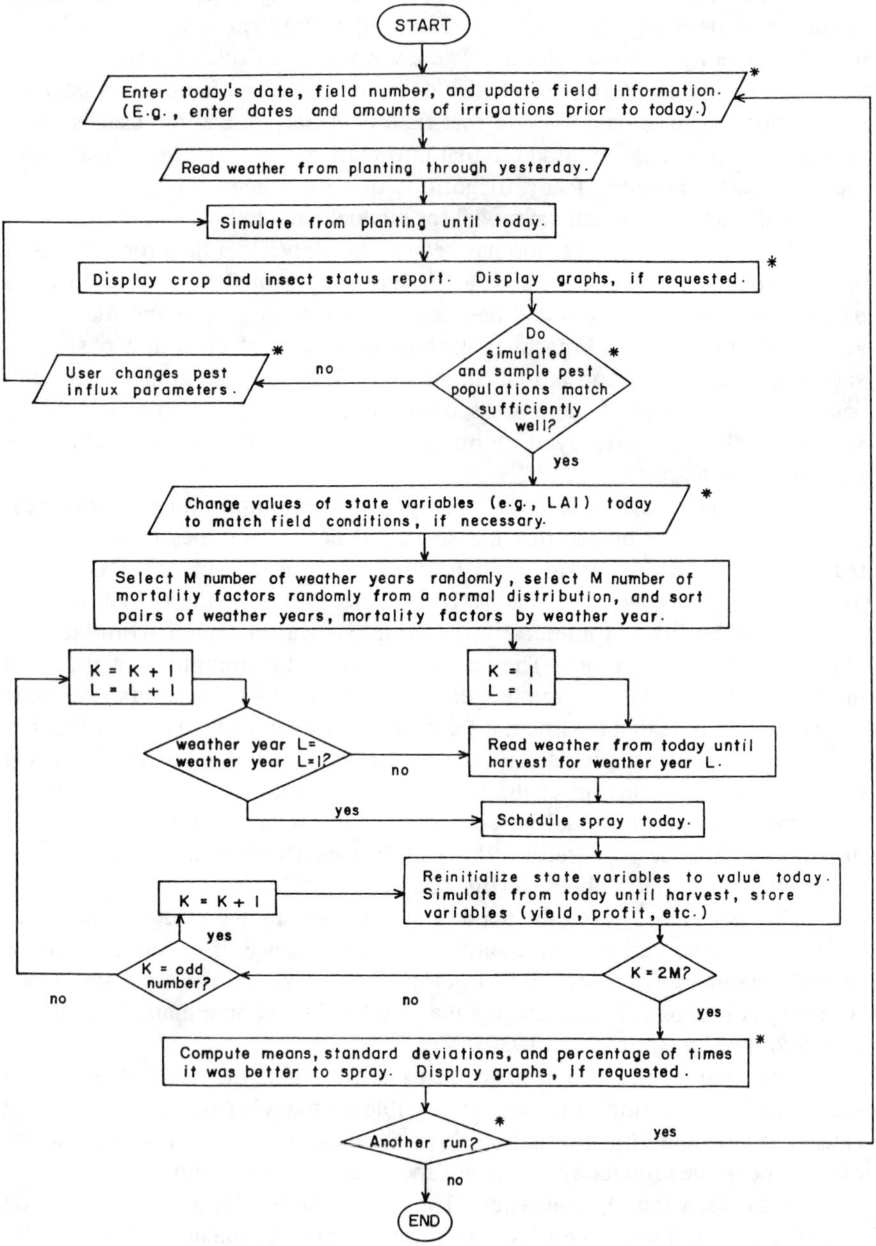

Figure 12.1–Flow chart of interactive, real-time pest management decision algorithm. The user interacts with the computer to make choices or to enter data at points in the algorithm marked by *. (Wilkerson et al., 1983b.)

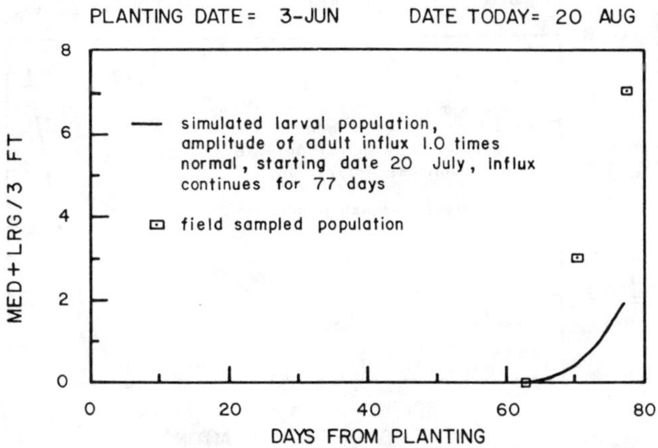

Figure 12.2–Comparison of field sampled medium + large VBC larval populations with simulated populations when adult influx parameters are set to those values which yield a "typical year" pattern of larval population. (Wilkerson et al., 1983b.)

soybean selling price, are known and that the deterministic crop and insect models accurately integrate the effects of weather and pests on yield. An open loop control problem can then be solved to determine the optimal amount and type of pesticide to apply at to and the optimal timing, amounts, and types to be applied for the rest of the year. Mathematically stated, the problem is:

$$\underset{u_{jt}}{\text{Max}}\, \pi\left(\overline{X}_{t_o}, u_{jt}\right) = P\ f_y\left(\overline{X}_{t_o}, U, W, \overline{I}\right) - \sum_{jt} c_j\, u_{jt} - CF \qquad t > t_o \qquad t < t_o \tag{12.1}$$

where

π = profit
u_{jt} = the amount of pesticide of type j to apply on day t
$\overline{X}_{t_o}$ = a vector of values of crop and pest state variables at time t_o
U = a matrix of the control variables, u_{jt}
W = a matrix of weather variables, e.g., temperature, rainfall each day
I = a vector of pest influxes through time
P = the price for the yield component $f_y(\cdot)$
$f_y(\cdot)$ = the yield of the crop expressed as a function of X_{t_o}, U, W, and I
c_j = the cost for pesticide j, including application cost
CF = the remaining cost of production

Because future weather and pest fluxes are uncertain, future crop and pest state variables are also uncertain. Thus, results from the deterministic problem cannot be used for a season. Assume that W and I are random variables and that

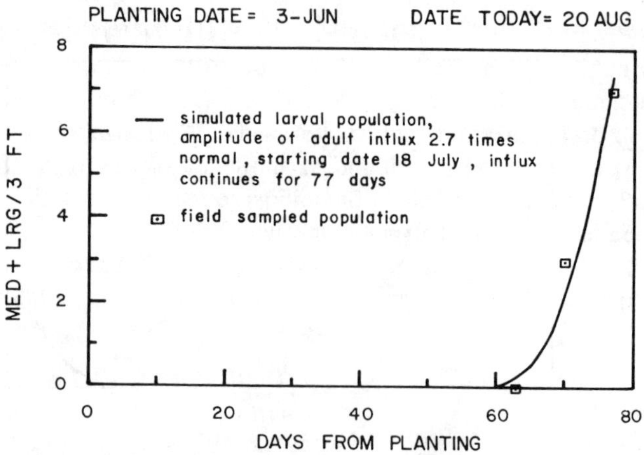

Figure 12.3–Comparison of field sampled medium + large VBC larval populations with simulated populations when adult influx parameters have been adjusted to yield a good fit. (Wilkerson et al., 1983b.)

F(W, I) describes their joint distribution. The problem now becomes one of maximizing the expected profit, in an open loop stochastic control problem:

$$\underset{u_{jt}}{\text{Max}}\ E\left[\pi_{t_o}\left(\overline{X}_{t_o},\ u_{jt},\ F(W,\ \bar{I})\right)\right]$$

$$t > t_o \qquad\qquad t < t_o \tag{12.2}$$

where E denotes expected value.

The control matrix U*, determined at time to to yield the maximum expected profit, may be different from the control matrix U**, determined at time t_o+k (where k is the scouting interval), since there is more information available at time t_o+k than there was at time t_o. Thus the optimization problem must be solved again on each decision date, using feedback information about the real state of the system at time t to obtain u^*_{jt}, the control(s) to be applied at time t. This is the process of stochastic open loop feedback control described by Zavaleta et al. (1980).

The stochastic open loop feedback control problem becomes very large when all combinations of u_{jt}, $t > t_o$, are considered. It was decided, therefore, to restrict the optimization problem to select only $u^*_{jt_o}$ subject to a background strategy (S) for $t > t_o$+k, i.e.,

$$\underset{u_{jt}}{\text{Max}}\ E\left[\pi_{t_o}\left(\overline{X}_{t_o}, u_{jt}, F(W, \bar{I}), S\right)\right]$$

$$t_o < \tag{12.3}$$

The strategy S is a set of rules that causes pesticide type j to be applied on any decision date after $t_0 + k$ when specified crop and insect conditions are met. For example, S may consist of a set of rules that causes methomyl (Lannate) to be sprayed at a certain rate whenever the medium plus large larval population exceeds 12 per 91 cm (3 ft) of row.

The decision model is therefore a conditional optimization problem which will result in a u^*_{jto} that depends on the selected strategy. However, it was assumed that if S is selected in such a way that it is a good or "optimal" long-term strategy to follow, then the algorithm will be very practical in tailoring an intended strategy to the actual conditions realized in the current season. Swaney et al. (1983) found that such an approach improved irrigation decisions.

For the simulations discussed in this chapter, the optimization problem was further restricted to only consider a subset of pesticides recommended by the extension service for control of velvetbean caterpillar, with the pesticides being applied at the minimum recommended rate. It was assumed that only one pesticide will be applied on any one day.

Simulation of Future Conditions. In order to incorporate the uncertainty about the future, W and I are assumed to be independent, and that $F(W, I) = F(W) \cdot F(I)$. Seventeen years of historical weather data were available for Gainesville, Florida. With this procedure, it was assumed that the probability of the weather during the remainder of this year being similar to any one of these past years, is identical. In other words, the occurrence of a weather year like year x is equally as probable as a weather year like year y.

In order to incorporate the uncertainty about future pest population levels in the field, daily mortality was the chosen variable rather than the pest influx. In the VBC model (Wilkerson et al., 1986), daily survival, S_i, is expressed as:

$$S_i = (1 - \mu_{ip})(1 - \mu_{iT})(1 - \mu_{iF})(1 - \mu_{iB}) \tag{12.4}$$

where

i = stage (1 = egg, 2 = small larva, etc.)
μ_{ip} = daily mortality due to predators
μ_{iT} = daily mortality due to pesticides which have been applied
μ_{iF} = daily mortality due to food shortage
μ_{iB} = background mortality due to factors other than predators, pesticides, or food shortage

In the deterministic model, μ_{iB} is assumed to be constant for each stage throughout the season. This assumption was modified in two different ways. First it was assumed that μ_{iB} may vary from field to field according to a normal distribution, i.e., for any particular simulation it will be constant, but its value will vary from run to run ($\mu_{iB} = h\mu_{iA}$, where h is from a normal distribution with mean 1.0 and variance 0.04). On the average, its value will be the value used in the deterministic model (μ_{iA}). A Monte Carlo technique is used to select weather years from a uniform distribution and to select a background mortality factor ($h\mu_{iA}$) from a normal distribution for each pair of spray/don't spray simulations. Calculation of expected net profit and variance of profit when pesticide k is applied at the recommended rate is according to the standard formulas for mean and variance of a random sample:

$$E\left(\pi_{t_o} \mid u_{kt_o}\right) = \sum_{m=1}^{M} \left[\frac{\bar{\pi}_m\left(X_{t_o}, u_{jt}, u_{kt_o}, W_m, h_m, S\right)}{M}\right] \qquad t < t_o \tag{12.5}$$

$$\mathrm{Var}\left(\pi_{t_o} \mid u_{kt_o}\right) = \sum_{m=1}^{M} \left(\frac{(\bar{\pi}_m - \pi)^2}{(M-1)}\right) \tag{12.6}$$

where π_m is the net profit when simulation from today until harvest uses weather W_m and mortality $h_m\mu_{iA}$, and M is the number of pairs of W_m and h_m simulated.

The second manner in which mortality was varied was to assume that the background mortality could take on only five different values ($\mu_{iB} = h_m\mu_{iA}$, where h_m = 0.6, 0.8, 1.0, 1.2, or 1.4 with probabilities, p, of 0.0668, 0.2417, 0.3830, 0.2417, and 0.0668, respectively). With h_m taking only five different values, the need to use Monte Carlo sampling of weather and mortality factors vanishes. If we again assume each weather year has an equal probability of occurring, then we can make 5N simulations (where N is the number of weather years) and determine the expected value and variance of profit directly for each decision path:

$$E\left(\pi_{t_o} \mid u_{kt_o}\right) = \sum_{n=1}^{N} \sum_{m=1}^{5} \frac{p_m}{N} \left[\pi_{mn}\left(X_{t_o}, u_{jt}, u_{kt_o}, W_n, h_m, S\right)\right] \qquad t < t_o \tag{12.7}$$

where π_{mn} is the simulated net profit if pesticide k is applied today and mortality factor h_m with probability p_m is assumed to apply for the remainder of the season. Historical weather data from year n is used in the simulation. The variance of profit $\mathrm{Var}(\pi_{t_o})$ can be calculated as:

$$\begin{aligned}\mathrm{Var}\left(\pi_{t_o} \mid u_{kt_o}\right) &= E\left(\pi_{t_o}^2 \mid u_{kt_o}\right) - \left[E\pi_{t_o} \mid u_{kt_o}\right]^2 \\ &= \sum_{n=1}^{N} \sum_{m=1}^{5} (p_m\pi_{mn}^2 / N - \left[\sum_{n=1}^{N} \sum_{m=1}^{5} (p_m\pi_{mn}) / N\right]^2 \end{aligned} \tag{12.8}$$

VBC Model Description

The model for Velvetbean Caterpillar chosen for PESTDEC was developed by Wilkerson et al. (1986). The following description of the VBC model is taken from that publication. The VBC populations are divided into six developmental stages: eggs, small larvae (0-12.5 mm long), medium larvae (12.5-25 mm long), large larvae (> 25 mm long), pupae, and adults. Larvae are separated by size rather than instar so that model results can be compared with field sample counts. The small size class corresponds roughly to larval instars 1 and 2; the medium to instars 3 and 4; and the large to instars ≥ 5.

The population model is an extension of the von Foerster (1959) model, in that an age structure (a) is maintained within each developmental stage (i). The change in numbers of individuals (N_i) in each stage can be expressed as:

$$\frac{\partial N_i}{\partial t} + \frac{\partial N_i}{\partial a} = -\mu_i N_i - T_i N_i + E_i \tag{12.9}$$

where μ_i is the stage-specific mortality, T_i is a developmental function, and E_i is the number that enter a stage, either immigrating to the field or developing from the previous stage. For small larvae to adult stages (i = 2, 6) E_i can be calculated as:

$$\begin{aligned} E_i(0, t) &= T_{i-1}N_{i-1} + I_i \\ E_i(a, t) &= 0 \quad a \neq 0 \end{aligned} \tag{12.10}$$

where I_i, the immigration to the field, is not zero only for adults (i = 6). For eggs (i = 1),

$$\begin{aligned} E_1(0, t) &= r_s \int_0^{25} \rho(a, t)\ N_6(a, t)\, da \\ E_1(a, t) &= 0 \quad a \neq 0 \end{aligned} \tag{12.11}$$

where r_s is the ratio of females to total number of adults (0.5), $\rho(a, t)$ is the number of eggs laid by a female of age a at time t, and 25 is the maximum age for egg-laying adults. The calculation of daily values for T_i, μ_i, I_i, and ρ are described in the following sections.

Development. The length of time required for VBC to develop from one stage to the next not only varies with temperature, but also varies between individuals raised at identical temperatures. Reid (1975), Leppla et al. (1977), and Johnson et al. (1983) have conducted experiments to determine the average length of time required for larval development at various constant temperatures. Johnson et al. (1983) reported mean time from oviposition to medium larval size, to large larval size, to pupal stage, and to adulthood for VBC raised at nine different constant temperatures. In addition, they reported the mean rate of development for each of these stages and the standard deviations for these rates.

This model incorporates the effect of varying temperature on development by calculating physiological days. Information reported by Johnson et al. (1983) was used to create a table for temperature and physiological days (table 12.1). For example, at 15.6° C, it takes ca. three-fold as long for an average VBC to go from oviposition to adult as it does at 26.7° C. Thus, one day at an average temperature of 15.6° C equals 0.33 physiological days. For this model a lower threshold for development of 13.0° C (Leppla et al., 1977) was used.

The variability among individuals in developmental rates has been accounted for in the model by using a cumulative distribution function taken from Stinner et al. (1975):

$$F(a_p) = (1 - Z)\,\Theta Z^k \tag{12.12}$$

where

$$Z = \frac{(B - a_p)}{(B - A)} \tag{12.13}$$

with a_p representing physiological age in our case, and Θ and k being constants. F is the proportion of individuals of the same age that have completed development by the time they have reached physiological age a_p; A is the number of days required for the first insect to complete development; and B is the number of days required for the last insect to complete development. The time period (B-A) represents the total time during which insects are completing development and passing to the next stage. Z, therefore, starts at 1.0 when the first insects complete development at a_p = A and drops to 0.0 as the last insects make their stage transitions at age a = B. Both Θ and k are empirical factors that determine the shape of the $F(a_p)$ curves as a varies from A to B. Stinner et al. (1975) found that a value for k of 2.8 worked well for a number of different insects, therefore k was set to 2.0. To calculate Θ, one must know A, B, and the mean time required. Johnson et al. (1983) presented a mean time of 29.3 days at 26.7° C, but they did not provide the range. It was assumed, as did Sharpe et al. (1981), that the range of development rate is the mean rate ± three standard deviations. This yielded a range at 26.7° C of 23.26-39.06 days. Johnson and his coworkers (1983) reported a mean development rate of 0.0343 with a standard deviation of 0.0029. A value of 1.886 was calculated for Θ.

A similar distribution function for development from one stage to the next (e.g., egg to small larva) was used as that used for the overall development from oviposition to adulthood. Using values for e and k of 1.886 and 2.0,

Table 12.1. Mean developmental times and physiological days for VBC held at various constant temperatures (Wilkerson et al., 1983b)

Temp (°C)	Mean Development Time from Oviposition to Adulthood Days*	Physiological Days per Calendar Days
13.0	–	0.00
15.6	90.0	0.33
18.3	71.2	0.41
21.1	42.1	0.70
23.9	33.1	0.88
26.7	29.3	1.00
29.4	25.8	1.14
32.2	20.8	1.41
35.0	22.7	1.29
37.8	22.7	1.29

* Data from Johnson et al. (1983).

respectively, for all stages and the mean duration of development time for each stage obtained from Johnson et al. (1983), ranges for each stage were determined (table 12.2). Since Johnson et al. (1983) did not determine duration of time from oviposition to small larva, a value of 3.2 days was estimated from Moscardi (1979) as the mean duration of the small larval stage at 26.7° C. Duration of the egg stage was then obtained by subtraction of the duration of the small larval stage from the average time from oviposition to medium larval stage. Leppla et al. (1977) reported that peak egg hatch occurred seven days post-oviposition at 26.7° C.

The proportion of those present at time t of age a, who complete stage i at time t, $T_i(a,t)$, could be obtained for use in equation 12.9 by differentiating the $F_i(a_p)$ cumulative distribution function obtained above.

$$T_i(a, t) = \frac{F_i'\left[a_p(a, t)\right]}{1 - F_i\left[a_p(a, t)\right]}\left[1 - \mu_i\right] \quad (12.14)$$

The term $[1 - \mu_i]$ is included in equation 12.14 since the cumulative distribution function is based on the number that survive the stage and develop to the next stage. The discrete representation of the derivative of F_i is:

$$F_i'\left[a_p(a, t)\right] = F_i\left[a_p(a, t)\right] - F_i\left[a_p(a-1, t-1)\right] \quad (12.15)$$

This form was used in the computer simulation.

Table 12.2. Estimated range of development time for all life stages of VBC at a constant temperature of 26.7° C* (Wilkerson et al., 1983b)

VBC Stage	Length of Development Time at 26.7° C (days)	Estimated No. of Days for First Individual to Complete Stage	Estimated No. of Days for Last Individual to Complete Stage
Egg	7.4	6.2	9.5
Small larva	3.2	2.0	5.1
Medium larva	3.8	2.6	5.7
Large larva	4.7	3.5	6.6
Pupa	10.2	9.0	12.2
Total	29.3	23.3	39.1

* Data from Johnson et al. (1983).

Oviposition. The total number of eggs laid by a female VBC in her lifetime is a function of temperature (Moscardi et al., 1981). The proportion of those eggs laid on any particular day (t) is a function of temperature T_A and adult age, a (Moscardi, 1979; Moscardi et al., 1981). Data presented by Moscardi (1979) were used to determine a relationship between physiological age (a_p) and proportion of total lifetime eggs laid on a particular day, $f_o(a, t)$:

$$f_o(a, t) = \begin{cases} 0.0 & a_p < 2 \\ c_1 a_p + c_2 & 2 \le a_p < 4 \\ c_3 + c_4 a_p + c_5 (a_p)^2 & a_p \ge 4 \end{cases} \quad (12.16)$$

where $c_1 - c_5$ are coefficients determined by regression (c_1 is 0.083, c_2 is –0.166, c_3 is 0.26, c_4 is –0.027, and c_5 is 7.2×10^{-4}; $R^2 = 0.87$ for $a_p > 4$).

Each day the total number of eggs (σ) a female would lay in her lifetime if the temperature remained constant at the temperature prevailing that day is calculated from a quadratic relationship derived from data reported by Moscardi (1979):

$$\sigma = \begin{cases} 0 & T_A < 18.5 \\ C_6 + c_7 T_A + c_8 T_A^2 & 18.5 \le T_A \le 33.9 \\ 0 & T_A > 33.9 \end{cases} \quad (12.17)$$

where c_6 is –9,988, c_7 is 8.33.95, and c_8 is –15.9 ($R^2 = 0.99$). This yields a maximum egg production of 945 eggs per female if temperature is optimal at

26.2° C. Number of eggs (p) laid by a female adult of age a at time t is then calculated for use in equation 12.11 as:

$$p(a, t) = \sigma f_o(a, t) \tag{12.18}$$

Mortality. At present, mortalities inflicted by predators, (μ_{iP}), insecticides (μ_{iT}), and food shortage (μ_{iF}) are incorporated into the model explicitly. Functionally, mortality is represented as:

$$\mu_i = 1 - S_i \tag{12.19}$$

where

$$S_i = (1 - \mu_{iB})(1 - \mu_{iF})(1 - \mu_{iP})(1 - \mu_{iT}) \tag{12.20}$$

and μ_{iB} represents mortality due to factors other than predators, insecticide, or food shortage. For the simulations reported in this chapter, μ_{iT} was always zero because no insecticides were applied to the fields used for calibration and validation of the model.

Background Mortality. For all stages except pupae, background (μ_{iB}) mortality was considered to be a constant proportion throughout the season: 0.10, 0.05, 0.05, 0.02, and 0.05 for eggs, small, medium and large larvae, and adults, respectively. These value were determined by calibration to the 1980 field data (Elvin, 1983). C. S. Barfield (personal communication) has found that pupal mortality is much higher early in the season before the canopy has shaded the ground than it is after canopy closure. Background mortality for pupae was made a function of leaf area index (L) generated by the soybean crop growth model (Wilkerson et al., 1983a) to which the VBC model is coupled. Daily mortality was estimated from Barfield's data to be 0.20 early in the season and 0.057 after canopy closure. Although he did not measure leaf area index, it was estimated to be 2.5 at the time the early season mortality calculations were made. Lacking information on intermediate points. A linear decrease in mortality between L of 2.5 and 3.835 (point at which ground cover reaches 100% in soybean growth model), was assumed. Therefore, the background mortality for pupae is calculated as:

$$\mu_{5B} = \begin{cases} C_9 & L < 2.5 \\ C_{10}L + C_{11} & 2.5 \le L \le 3.835 \\ C_{12} & L > 3.835 \end{cases} \tag{12.21}$$

where C_9 is 0.20, C_{10} is –0.107, C_{11} is 0.47, and C_{12} is 0.057. Ground cover is assume to reach 100% when L is 3.835.

Mortality Due to Food Shortage. The mortality factor due to a shortage of food (μ_{1F}) is equal to zero unless leaf areas has been zero for two days or more. This factor increases linearly from zero to 1.0 between 2 and 10 days after leaf area becomes zero. This relationship was assumed to reflect the decline in larval populations to zero within two weeks of 100% defoliation (Elvin, 1983).

Mortality Due to Predators. Many species of arthropods have been identified as predators of VBC (Buschman et al., 1977a; Elvin et al., 1983). The information to model each of these species and its effects on VBC population numbers individually is not available, however the assumption incorporated into the model is to make mortality due to predators a function of the total number of predators present in the field per 91-cm sample. At present, predator population density is a function of Julian date of the year, if the field has not been sprayed. Field samples taken during 1980 and 1981 in unsprayed fields were averaged to obtain the numbers used in the model (Elvin, 1983; Elvin et al., 1983). A similar seasonal pattern of predator populations was reported by Marston et al. (1979).

Each day a predator index is calculated from the total number of predators present as follows:

$$PI = \min\left[1.0, \frac{PR\,(t)}{PR_{max}}\right] \qquad (12.22)$$

where PR(t) is the number of predators present today and PR_{max} is the number of predators present when VBC mortality due to predators is at a maximum. PR_{max} has been set to 8.0 in the model so that PI is 1.0 if there are > 8.0 predators per 91 cm of row.

Mortality of all stages of VBC due to predators is calculated as a linear function of PI:

$$\mu_{iP} = k_i\,PI \qquad (12.23)$$

where k_i is a constant specific to each stage (0.19, 0.13, 0.07, 0.05, 0.0, and 0.0 for stages egg, small larva, medium larva, large larva, pupa, and adult, respectively). The values of PR_{max} and k_i for all stages were determined by calibration of simulated results to the 1980 field data set (Elvin, 1983).

Immigration. The model is initialized by the influx of one-day-old adults into the field. VBC do not overwinter in north Florida (Buschman et al., 1977b). Even though adults begin to appear in flight traps in June or July, larvae do not show up in soybean field samples usually until August or September. The number of larvae that occur in unsprayed fields varies greatly from year to year, field to field, and the factors that influence this variability are at present unknown. The simulation began each year by assuming a daily influx of adults. Weekly averages of nightly flight trap catches for Quincy, Florida, in 1978 were used to construct the first half of our standard adult influx curve (Herzog and Todd, 1980). Linearly interpolated daily influx values were computed from the weekly values. The problem of converting flight trap numbers to numbers per square meter was resolved by multiplying the numbers reported by Herzog and Todd (1980) by different factors until the simulated larval population peaks matched those seen in unsprayed fields in an average year (ca. 12 large larvae per 91 cm of row). The factor that worked best was 3.36×10^3. It was not possible to use the values reported by Herzog and Todd (1980) at the end of the season when flight traps were catching adults emerging within the field as well as those entering the field. The second half of the curve was obtained, therefore, by calibration of simulated larval counts to the 1980 field counts (Elvin, 1983).

As part of PESTDEC the standard influx curve may be modified in any of three ways to fit a particular field: 1) the starting date of the influx can be changed; 2) the amplitude of the influx can be changed by multiplying each day's influx number by a constant; and 3) the number of days over which the influx occurs can be increased or decreased. The first three or four nonzero small larval field counts were used to estimate the actual beginning influx date for a particular field and the amplitude of the influx. For the simulation presented in this chapter the standard values were chosen to be two weeks early, on time, two weeks late, and 0.5 times normal population, normal population, and 2.0 times normal population for the starting date and amplitude adjustments, respectively.

Procedure

To evaluate the performance of PESTDEC three weather years of Gainesville, Florida from 1978, 1979, and 1980 were chosen. The year 1978 was a very bad weather year for crop production with a severe drought, and the year 1980 had an early water stress period, with rainfall abundant during the later part of the season. In 1979 the rainfall was excellent. Using these data for weather, a variety of pest conditions were simulated. Table 12.3 presents the descriptions of each set of simulations. The inputs varied include irrigation, pest influx, and pesticide cost. The model would automatically irrigate following a strategy to avoid most water stress under Florida conditions and for the no-irrigation simulations only rainfall provided water for crop growth. Pest influx patterns were varied by increasing the influx populations from normal, one half normal, and twice normal. In addition, the timing of the influx was adjusted from the normal time of 21 July to 7 July for the early influx and to 4 August for the late influx. The crop planting date was selected to be 3 June using Bragg Soybeans. To vary the cost of pesticide the pesticide material cost and application cost were varied from 100, 200, and 400% with values of \$11.66, \$23.33, and \$46.65/ ha, respectively, assuming the pesticide lannate at 0.7 l/ha active ingredient.

The optimal values were obtained by simulating all possible spray dates from 1 August to 26 September of each year with only one spray allowed for each seventh day scouting date. The optimal value was chosen as the spray schedule that produced the highest net profit. The current extension recommended VBC control strategy is to apply a pesticide, when scouting

Table 12.3. List of simulations made using SICM and PESTDEC (Table 2 from Wilkerson and Mishoe, 1988)

Item	Planting Date	Irrigation	VBC Population Influx Size	Time	Pesticide Cost
19xxA	3 June	no	average	normal	100%
19xxB	3 June	no	low	normal	100%
19xxC	3 June	no	high	normal	100%
19xxD	3 June	no	average	early	100%
19xxE	3 June	no	average	late	100%
19xxF	3 June	yes	average	normal	100%
19xxG	3 June	yes	low	normal	100%
19xxH	3 June	yes	high	normal	100%
19xxI	3 June	no	average	normal	200%
19xxJ	3 June	no	average	normal	400%
19xxK	17 June	no	average	normal	100%
19xxL	20 May	no	average	normal	100%

indicates 12 medium plus large larvae per 91 cm (3 ft) of row. This strategy was used to determine the yield and profit and presented in the column labelled Extension. As a possible new strategy, a threshold of 6 medium plus large larvae was simulated and presented in the 1/2 Extension labeled column in a following table.

The expected damage strategy was an attempt to develop a novel decision criteria. PESTDEC was used to simulate until today's date, then the model simulated to the end of the season first assuming no additional damage from insect to the crop, and second assuming damage for the insect in place with no additional influx of pest. The cost of the pesticide application was compared to the potential damage from the insect and if the cost of applying the pesticide was less than the crop loss, then a spray was scheduled. In the last procedure used, labeled PESTDEC, decisions were made using the framework presented earlier in the model description where future conditions were simulated with various insect mortalities and weather years and a decision made based upon the maximum expected net profit.

Results and Discussion

Table 12.4 presents the results of the simulations conducted using SICM and PESTDEC. The average net profit for the optimal pest control schedule was $293, however, none of the procedures achieved the optimal decisions in all cases. It was surprising that all of the procedures performed well for these conditions, however, the cases presented here are in fact very limited. The overall best procedure was the PESTDEC model, even though it was only slightly better than the threshold of six larva per 91 cm of row and the damage assessment technique. The procedure of using a threshold of 12 medium plus large larva per 91 cm of row performed well except for the 1980 simulations assuming no irrigation. During that year the canopy of the soybeans did not develop at its normal rate due to the water stress during the early growing

Table 12.4. Comparison of simulated net profit for each set of inputs listed in Table 12.3 (Table 3 for Wilkerson and Mishoe, 1988)

	Net Profit ($ / ha)				
Item	Optimal	Extension	1 / 2 Extension	Damage Cost	PESTDEC
1978A	−128	−129	−138	−129	−129
1978F	315	310	315	308	308
1978I	−128	−141	−161	−142	−142
1978J	−128	−164	−208	−166	−166
1979A	541	530	541	541	541
1979B	546	542	546	546	546
1979C	530	530	530	530	530
1979D	528	520	524	513	515
1979E	543	538	538	538	538
1979I	518	518	517	518	518
1979J	495	495	471	495	495
1979K	469	456	469	464	469
1979L	575	575	568	575	575
1980A	201	166	200	201	200
1980B	212	176	202	212	212
1980C	190	174	174	188	188
1980D	199	133	188	195	199
1980E	205	178	178	204	204
1980F	428	408	428	421	428
1980G	433	429	433	433	433
1980H	418	413	413	406	418
1980I	177	155	177	177	177
1980J	131	131	130	130	130
1980K	186	124	181	186	181
1980L	162	141	161	162	162
1981A	165	159	165	155	165
1982A	392	381	392	392	392
1984A	16	10	4	10	16

season. Leaf area became more valuable to the crop and the threshold failed to provide close to optimal decisions. This suggests one valuable use of the integrated models is that they can be used to evaluate strategies for pest control for use independent of the model.

The cost of the tactic (pesticide) and the value of the crop determines the value of choosing an optimal control decision. This can be seen in the I and J simulations where the cost of the pesticide was increased. The potential for loss greatly increases. The case of Lannate applied to soybeans represents a low cost tactic on a low valued crop. On other crops such as vegetables the small changes in profit reflected here can greatly increase in value.

A major problem with all of the decision making procedures is the uncertainty of future conditions. With PESTDEC a range of expected conditions are examined in an attempt to reduce the chance of being totally wrong. The threshold approach does not consider future events. By having the expected range of possible outputs the user can also access the risk associated with the decision. A risk averse producer may choose a tactic that reduces the risk of profit loss.

A major limitation of all of the methods examined here is that only a component of the system is considered. At least the PESTDEC approach evaluates the impact to the crop within the current and expected future environmental conditions. Planting date, crop phenology, other pest, etc., are part of the SICM model, however, they have not been evaluated within the scope of this study. PESTDEC does not allow for the consideration of multipest in making a decision. There are, of course, many other conditions that are not included at this point. It is in fact not recommended that this approach should be extended to include more components of the production system directly. However the model can be used to generate relatively simple sets of rules that can help select the possible control options based upon the general state of the production system. These rules can be implemented in an Expert System format with the models being used to evaluate various problem conditions that cannot be reduced to relatively simple rules.

Conclusions

The framework presented is a promising technique that can be used by producers to improve their profit by improving pest control procedures, however, the model should not be used without adequate evaluation of all information available. It is very difficult to field test a model of this type, however, some effort in this direction is needed. Using a tested model to determine new and improved control strategies is a major use of these models. Expert systems represent a way of managing the entire production system and allowing the model to evaluate that part of the system that can be simulated.

References

Arkin, G. F., S. J. Maas and C. W. Richardson. 1980. Forecasting grain sorghum yields using simulated weather data and updating techniques. *Transactions of the ASAE* 23(3):676-680.

Arkin, G. F., and W. A. Dugas. 1981. Making weather and climate-dependent crop management decisions. In *Computer Techniques and Meteorological Data Applied to Problems of Agriculture and Forestry: A Workshop*, ed. A. Weiss, 223-237. American Meteorological Society.

Buschman, L. L., W. H. Whitcomb, R. C. Hemenway, D. L. Mays, R. Ru, N. C. Leppla and B. J. Smittle. 1977a. Predators of velvetbean caterpillar eggs in Florida soybeans. *Environ. Entomol.* 6:403-407.

Buschman, L. L., W. H. Whitcomb, T. M. Neal and D. L. Mays. 1977b. Winter survival and hosts of the velvetbean caterpillar in Florida. *Florida Entomology* 60:267-273.

Elvin, M. K. 1983. Quantitative estimation of rates of arthropod predation on velvetbean caterpillar (Anticarsia gemmatalis Hubner) eggs and larvae in soybeans. Ph.D. diss. University of Florida, Gainesville.

Elvin, M. K., J. L. Stimac and W. H. Whitcomb. 1983. Estimating rates of arthropod predation on velvetbean caterpillar larvae in soybeans. *Florida Entomology* 66:319-330.

Herzog, D. C. and J. W. Todd. 1980. Sampling velvetbean caterpillar on soybean. In *Sampling Methods in Soybean Entomology*, ed. M. Kogan and D. C. Herzog. New York: Springer.

Johnson, D. W., C. S. Barfield and G. E. Allen. 1983. Temperature dependent developmental model for the velvetbean caterpillar (Lepidoptera: Noctuidae). *Environmental Entomology* 12:1657-1663.

Leppla, N. C., T. R. Ashley, R. H. Guys and G. D. Butler. 1977. Laboratory life history of the velvetbean caterpillar. *Annals of the Entomological Society of America* 70:217-220.

Marston, N. L., G. D. Thomas, C. M. Ignoffo, M. R. Gebhardt, D. L. Hostetter and W. A. Dickerson. 1979. Seasonal cycles of soybean arthropods in Missouri: Effect of pesticidal and cultural practices. *Environmental Entomology* 8:165-173.

McClendon, R. W., K. S. Akbay, L. G. Brown and J. W. Jones. 1979. Interactive models of the cotton crop and insect pests. ASAE Paper No. 79-5017. St. Joseph, MI: ASAE.

Meyer, G. E. and R. B. Curry. 1981. SOYMOD as a management tool for soybean production. ASAE Paper No. 81-4012. St. Joseph, MI: ASAE.

Mishoe, J. W., J. W. Jones, D. P. Swaney and G. G. Wilkerson. 1984. Using crop and pest models for management applications. *Agricultural Systems* 15:153-170.

Moscardi, F. 1979. Effect of soybean crop phenology on development, leaf consumption, and oviposition of *Anticarsia Remmatalis* Hübner. Ph.D. diss. University of Florida, Gainesville.

Moscardi, F., C. S. Barfield and G. E. Allen. 1981. Effects of temperature on adult velvetbean caterpillar oviposition, egg hatch, and longevity. *Annals of the Entomological Society of America* 74:167-171.

Reid, J. C. 1975. Larval development and consumption of soybean foliage by the velvetbean caterpillar, *Anticarsia Remmatalis* Hübner (Lepidoptera: Noctuidae), in the laboratory. Ph.D. diss. University of Florida, Gainesville.

Rudd, W. G. 1975. Population modeling for pest management studies. *Mathematics in Bioscience* 26:283-302.

Sharpe, P. J. H., R. M. Schoolfield and G. D. Butler, Jr. 1981. Distribution model of *Heliothis zea* (Lepidoptera: Noctuidae) development times. *Canadian Entomology* 113:845-856.

Stinner, R. E., G. D. Butler, Jr., J. S. Bacheler and C. Tuttle. 1975. Simulation of temperature-dependent development in population dynamics models. *Canadian Entomology* 107:1167-1174.

Swaney, D. P., J. W. Jones, W. G. Boggess, G. G. Wilkerson and J. W. Mishoe. 1983. Real-time irrigation decision analysis using simulation. *Transactions of the ASAE* 26(2):562-568.

Wilkerson, G. G., J. W. Jones, K. J. Boote, K. T. Ingram and J. W. Mishoe. 1983a. Modeling soybean growth for crop management. *Transactions of the ASAE* 26(1):63-73.

Wilkerson, G. G., J. W. Mishoe, J. W. Jones, W. G. Boggess and D. P. Swaney. 1983b. Within-season decision making for pest control in soybeans. ASAE Paper No. 83-4044. St. Joseph, MI: ASAE.

Wilkerson, G. G., J. W. Mishoe and J. L. Stimac. 1986. Modeling velvetbean caterpillar (Lepidoptera: Noctuidae) populations in soybeans. *Environmental Entomology* 15(4):809-816.

Wilkerson, G. G. and J. W. Mishoe. 1990. Pestdec: A simulation approach to insecticide application decision-making for soybeans. *Agricultural Systems* 32:143-157.

Zavaleta, L. R., R. D. Lacewell and C. R. Taylor. 1980. Open-loop stocchastic control of grain sorghum irrigation levels and timing. *American Journal of Agricultural Economics* 62:785-792.

13

Application of CIM for Control Decisions on *Heliothis*

Gordon L. Andrews and Larry G. Brown

Introduction

These thoughts of modeling and agriculture should be kept in mind to understand the rational for these trials. Modelers tend to be very good scientists and models have played a major role in the injection of more science into agriculture. If computerized models are to be utilized by agriculture, modelers must not only look to their scientific colleagues for critical evaluation but they must look to the agricultural community to learn how models can improve present technology or how present technology will modify their models. A model which does not include all variables can be utilized by agriculture if it can be used by the producer to improve his overall management scheme. The producer must be aware of the missing variables if he is to utilize such a model. The "Cotton and Insect Management (CIM) Model" is composed of a plant model COTCROP (Brown et al., 1985), insect growth models, and appropriate subroutines which allow for interaction between the crop, insects, and insecticides (Brown et al., 1983). The model was designed to study insect control strategies in cotton. It has produced new strategies for *Heliothis* in Mississippi (Brown and McClendon, 1982) which have altered the present control recommendations in Mississippi after these strategies were field tested. The trials in this chapter represent the first attempt to use the CIM model in real time decision making mode.

The major emphasis of these field trials was to further develop the "Cotton and Insect Management (CIM) Model" for use as a production tool. Some modification of the model was made before this trial was undertaken and further knowledge was gained during the trial to help modify this model and future models which are to be used to make production decisions.

Evaluation was also an objective; however, it was known from the outset of the trials that evaluation would be incomplete. At the time these trials were conducted the model was simply a research model which was converted to be utilized for these particular trials on only one computer. This limited the trial to only one site and this fact limited the evaluation from the outset. A great deal of labor was needed to collect and compile necessary field data so that decisions could be made and communicated to the producer, so that the insecticide applications could be made on time. As much data as was physically possible with the time and labor available was collected from the fields. As the trials progressed data collection beyond what had initially been planned was done. In spite of our best efforts, in retrospect, data gaps exist which prevent a completely quantitative evaluation. The evaluation presented in this chapter will be, in part, form data collected during the trial, observations made in the field by the authors, and observations made by the producer of the crop which were communicated to the authors. Experiments conducted as late as the 1986 growing season on aphids have influenced the opinion of the authors about the outcome of these trials.

Methods and Materials

A detailed description of methods and materials have already been presented (Andrews et al., 1984), so only those methods and materials essential for understanding will be offered at this time. The trials were carried out in 1983 and 1984 on two, 8.1 ha (20 ac) fields planted with Stoneville 825 variety cotton. The fields were selected in an attempt to isolate *Heliothis* as the only insect which would need management. Other insects, if any, would be managed on both fields equally with blanket applications. In 1983 *Heliothis* populations in field 1 were managed using current Mississippi Cooperative Extension Service recommendations (Anonymous, 1983-84) or current insect control (CIC). In field 2, *Heliothis* were managed according to the data generated by the simulation runs of the CIM model or computer assisted management (CAM). In 1984 the same two fields were used but CAM was imposed on field 1 and CIC on field 2.

To make decisions with the CIM model it was modified so that it would run using current weather conditions to the last available weather date and then switch to historic weather for the remainder of the season. The CIM model contained an insecticide spray routine which was based on data collected on the effects of EPN-methyl parathion on *Heliothis* (McDaniel, 1976) and beneficial insect populations (Andrews, unpublished). In 1983 no such data were available on the effects of pyrethroid insecticides on *Heliothis* and beneficial populations. The insecticide spray routine was modified for these trials using field observations (Andrews, personnel communication). Since the trials the spray routine has been updated with experimental data (Luttrell, unpublished). Non-reoccurring data such as soil variables, planting date, expected price of cotton, and cost of insecticide were first entered into the model at the start of the season.

Several inputs of reoccurring data were needed at varying intervals throughout the season. The CIM model requires daily weather information. Solar radiation and pan evaporation data from Stoneville, Mississippi, which is approximately 80.5 km (50 m) west of the fields was used. Rainfall and air temperature (maximum and minimum) data were collected adjacent to the field. Once weekly plant data along with pest and beneficial insect data were collected. Pest insects sampled were boll weevil, bollworm, and tarnished plant bug. Numbers of beneficial insects which the CIM requires were estimated using sweepnet samples. The plant information was utilized to assure that the crop generated by the model was similar to that of the field.

The *Heliothis* data utilized to initialize the model for the decision making process was percent eggs and larvae collected twice weekly using common scouting techniques. This scouting data included sufficient information on other insect populations so that blanket insecticide applications could be made on both strategies to eliminate any damage which may interfere with evaluation of *Heliothis* control.

Once the *Heliothis* population was initialized, the current weather file was up dated, estimated beneficial populations input, and fruit curves generated by the crop model were checked against field collected data. The decision runs could be initiated. The model was run utilizing current weather up to the present decision date then switching to historic weather one day after the last current weather file entry. The insect control strategy imposed on the simulation the day after the decision date and for the remainder of the season was the dynamic threshold described by Brown and McClendon (1983). Two simulated crops were produced with each historic weather year. The first simulation was with an insecticide application the day after the date of simulation and the second simulation did not include this application. The difference (first simulation minus the second simulation) in net returns of the two simulated crops was plotted along with 6 to 20 such differences obtained by simulating different historic weather years. If the mean difference was positive and greater than the cost of an insecticide application, the producer was then called and an application was made on the field designated as the CAM strategy.

The fields were picked and ginned separately by the producer who furnished yield data. In 1983 the plots were picked on 14 October and on 16 November. The 1983 harvest was nearer the normal harvest situation. In 1984 the plots were picked only one time on 27 November. Rains in October of 1984 prevented timely harvest.

Results and Discussion

The single objective of the trials was to manage *Heliothis* populations. To evaluate how well the CAM managed *Heliothis* populations, several variables were measured and compared to those of CIC.

The total amount of insecticide and total number of applications are probably good variables to start with to detect differences in *Heliothis* management strategies. In 1983 the CIC used five applications containing a total of 0.67 kg/ha (0.6 lb/ac) of fenvalerate to control *Heliothis* populations on the test plot and CAM used exactly the same number of applications and containing the same amount of fenvalerate. In 1984 CIC used one application containing 0.11 kg/ha (0.1 lb/ac) of fenvalerate and CAM used exactly the same number of applications containing the same amount of fenvalerate. The fact that both strategies utilized the same number and amount of insecticide

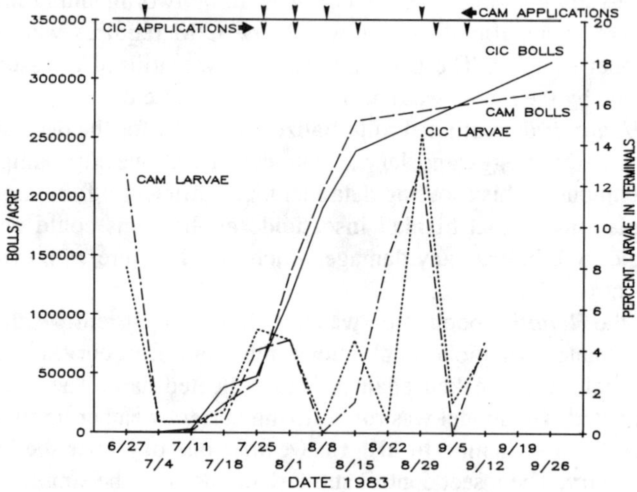

Figure 13.1–Mean bolls per hectare, percent *Heliothis* larvae in terminals, and timing of insecticide applications from fields managed with current insect control recommendations (CIC) and computer assisted management (CAM) during the 1983 cotton growing season.

seem almost coincidental when the application dates are examined. Of the six applications used by both strategies in the two years of the trials only two of the applications were applied to both CIC and CAM on the same dates (figs. 13.1 and 13.2).

The only difference seen between the two control strategies was in the timing of the applications as seen in figures 13.1 and 13.2. The CAM strategy started before the CIC and quit spraying before CIC in both trial years. Several other bits of information should be noticed and kept in mind about the two cotton growing seasons. The *Heliothis* larval counts are presented as larvae in terminals, which means that they are probably first, second, or third instar larvae. Subsequently boll damage from larvae found in terminals occurs approximately a week after the larvae are detected. Boll counts taken at the end of September or the first of October closely represent the bolls which are actually harvested. Using a little linear extrapolation and measuring on figures 13.1 and 13.2, one would estimate that most of the bolls harvested in 1983 were set prior to 20 August and in 1984 prior to 15 August.

The best way to determine if this difference in timing increased or decreased *Heliothis* control is to look at the object you are trying to protect, which is bolls. *Heliothis* damaged bolls in 1983 (fig. 13.3) cannot be statistically considered different on any date. However, collectively CIC had lower boll damage counts four of the six times measured in 1983 which gives CIC the edge in 1983. In 1984 (fig. 13.4) just the opposite was true and CAM had lower damage boll counts five of the six times measured. The differences between the two control strategies and the overall damage to bolls in most cases would be less than 11.2 kg of cotton/ha (10 lb/ac) using a mean of 136 bolls/kg

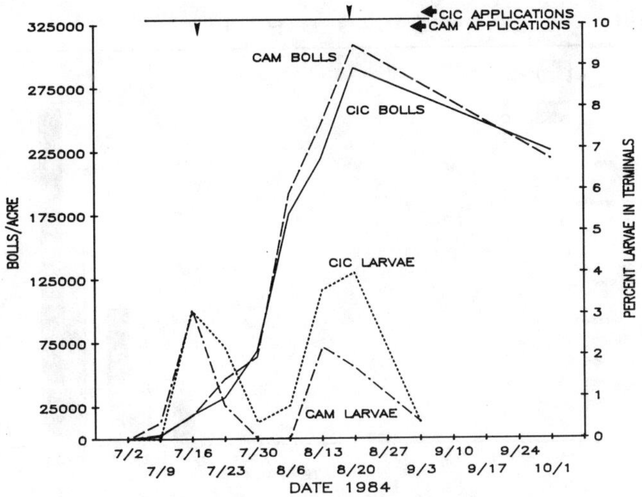

Figure 13.2–Mean bolls per hectare, percent *Heliothis* larvae in terminals, and timing of insecticide applications from fields managed with current insect control recommendations (CIC) and computer assisted management (CAM) during the 1984 cotton growing season.

(300 bolls/lb) of lint cotton. The mean bolls/kg of cotton can be as few as 114 (250/lb) early season and as high as 227 (500/lb) late season; therefore differences seen early season may be greater than the same difference late season.

After studying figures 13.1-13.4 one can only conclude that both strategies controlled *Heliothis* very well. Differences did exist in the timing of

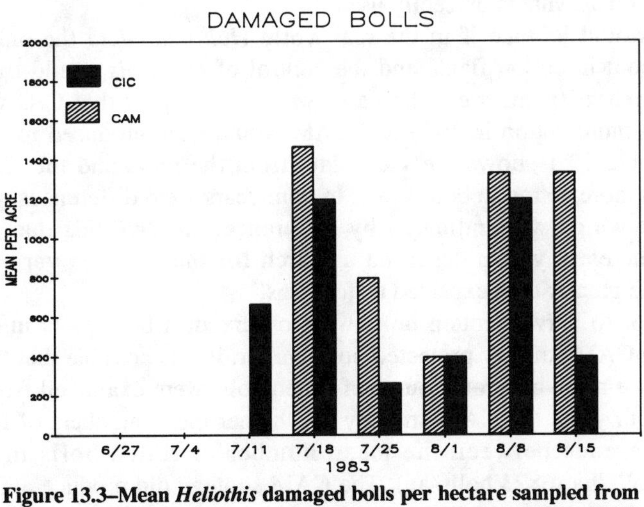

Figure 13.3–Mean *Heliothis* damaged bolls per hectare sampled from fields managed with current insect control recommendations (CIC) and computer assisted management (CAM) during the 1983 cotton growing season.

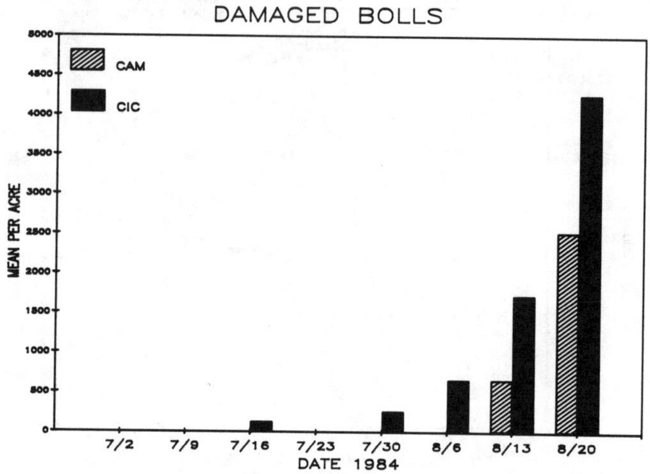

Figure 13.4–Mean *Heliothis* damaged bolls per hectare sampled from fields managed with current insect control recommendations (CIC) and computer assisted management (CAM) during the 1984 cotton growing season.

applications aimed at *Heliothis*. The first year, 1983, when some problems were encountered in modifying the simulated crop to match the crop in the field and fewer historical weather years were used, the difference in timing did not seem to help the CAM strategy protect bolls. In 1984 when fewer internal model errors were made the difference in timing seemed to produce differences in boll damage. The differences need to be looked into further. The modeling community should be proud of the fact that the first attempt at *Heliothis* control was as good as what is presently used.

It would be nice if in the real world *Heliothis* were the only insect to worry about in cotton fields and the control of *Heliothis* would translate into yield increase. If this were the case, we would expect that CIC would have produced more cotton in 1983 and CAM would have produced more cotton in 1984. Table 13.1 shows the yield data from the plots and the CIC strategy produced more cotton in both years. In both years yield differences were greater than that which was indicated by differences in *Heliothis* damage. These discrepancies in yields deserved a search for the possible variables which caused the greater than expected differences.

Prior to harvest rotten bolls were observed in both plots in both years. Since the CAM strategy protected bolls earlier, it was possible that this strategy may induce more boll rot. Counts of rotten bolls were examined from the plots and in both years the CAM strategy had higher mean numbers of rotten bolls. The difference between the mean number of rotten bolls in 1983 was 16,847 bolls/ha (6823 bolls/ac). The CAM strategy did produce a higher mean rotten boll count but the differences in rotten bolls between the two strategies was not near great enough to have produced all the yield difference between the

Table 13.1. Yields in kilograms of lint cotton per hectare from fields managed with current insect control recommendations (CIC) and computer assisted management (CAM) during the 1983 and 1984 cotton growing seasons and the difference in yield from the two fields (CIC-CAM)

Harvest Date	CIC	CAM	CIC-CAM
1983			
14 October	836	785	51
16 November	192	169	23
Total	1027	954	73
1984			
27 November	842	707	135

strategies. Through observation and soil testing, the field that the CIC strategy was imposed on in 1983 was better drained and the soil contained a high percent sand. Less boll rot, better soil, and possibly slightly better *Heliothis* control or some combination of these variables could sufficiently explain the increase in the CIC field in 1983.

In 1984 the CAM strategy was imposed on the same field as the CIC was imposed on in 1983 which was proclaimed in 1983 as being the field which was more favorable for cotton production. In 1984 the data collected (fig. 13.4) on worm damaged bolls showed there was a high probability of superior *Heliothis* control by the CAM strategy. The problem was then to explain how the CAM field in 1984 with all of its advantages could produce 135.5 kg of lint cotton/ha (121 lb/ac) less than the CIC field. Rotten boll counts showed that the CAM strategy produced a mean of only 5383 more rotten bolls/ha (2180 bolls/ac) than the CIC strategy. This small difference in boll rot could not explain the difference in yields of more than 135.5 kg (121 lb/ac) of lint cotton/ha, since the CAM strategy was on the field with higher production capabilities in 1984. In 1984 the application of fenvalerate to the CAM field in the middle of July did cause an increase in aphids in the CAM field. The aphid population was not considered to be high enough to lower yields sufficiently to off set the cost of an insecticide application by the entomologist (G. L. Andrews) in the field. Replicated tests conducted in the summer of 1986 on aphids showed that aphid populations (which were observed to be lighter than the population in the CAM plot in 1984) allowed to remain on cotton for the last three weeks in July and the first week in August reduced yields approximately 112 kg (100 lb/ac) of lint cotton/ha (Andrews, unpublished). Mid-season aphid populations in cotton are a rather recent phenomenon which can be initiated by the application of synthetic pyrethroid insecticides. At this point, information from literature about mid-season aphid populations and their effect on cotton yields is not available. The tests conducted in 1986 indicate the aphid population in the CAM plot of 1984 may have been responsible for much of the yield decrease in this plot.

This was the first attempt to use the CIM model as a production tool and hopefully this will lead to further modification of this model as a production tool. There are other ways to utilize this model and other simulation models to aid in cotton production. Further development of the CIM model as a production tool was an objective of this trial and some of the experience gained may be applicable to future attempts to utilize models to assist with insect management.

The most obvious fact which this trial points out is that models can not interact with variables which are not in the model. In some way models which will be used for production decisions must be flexible enough to interface with changes in the production system as quickly as possible. In this trial techniques and results were altered by using an insecticide which had been introduced into production agriculture for only five years. At the time this trial was conducted the effects of fenvalerate on the target pest had not been quantified and its interaction with non-target insects were not understood. Changes in agriculture such as new insecticides; resistance in *Heliothis* and other pests; and their interaction with beneficial arthropods will continue. Addressing the data needs of models as well as making models more flexible to accept such data could help modify models more quickly for production purposes.

In both years the simulated crop started to square sooner than the crop in the field for some unknown reason. Herbicide and seedling disease effects have been offered as possible variables which are not in the model but which could have caused the delay seen in the field. It was necessary to adjust the crop to match the field data. After this time no further adjustments were necessary. Some field data will be needed from fields where CAM attempts are made in the future to assure that insects are managed on a crop which is sufficiently similar to reality.

In these trials scouting data were used to make decisions. However, no quantitative method exists to convert percent eggs and larvae in terminals to eggs and larvae per unit area. In this trial the fraction of the percent eggs and larvae was multiplied by the plant population. While this seemed to agree with the whole plant counts, both the scouting sample and the whole plant sample had coefficients of variation greater than 50 near the 3704 larvae/ha (1500/ac) samples. This variation occurred in spite of the fact that the sample size used in this trial was greater than two times what is presently recommended. This points out the need for methods to convert present scouting data or new scouting methods for use with CAM. One must keep in mind that scouting cost, money and the accuracy of the data must be balanced with cost of obtaining the data.

Conclusions

This trial demonstrated that simulation models can be used to make control decisions on *Heliothis* which were as good as the currently used strategies. Differences between the CIC and CAM on *Heliothis* control caused differences in aphid populations which seem to have caused yield decreases in the CAM field in 1984.

References

Andrews, G. L., L. G. Brown and K. S. Akbay. 1984. Computer assisted cotton insect management. In *Proceedings of the Beltwide Cotton Production Research Conference*, 239-244. Memphis, TN: National Cotton Council.

Anonymous. 1983. Cotton insect control guide. Pub. No. 343. Mississippi Cooperative Extension Service, Mississippi State.

Anonymous. 1984. Cotton insect control guide. Pub. No. 343. Mississippi Cooperative Extension Service, Mississippi State.

Brown, L. G. and R. W. McClendon. 1982. A new *Heliothis* spp. control strategy for the Mississippi delta. In *Proceedings of the Beltwide Cotton Production Research Conference*, 191-193. Memphis, TN: National Cotton Council.

Brown, L. G., R. W. McClendon and J. W. Jones. 1983. Cotton and insect management simulation model, Chapter 17. In Agricultural Handbook No. 589: *Cotton Insect Management with Special Reference to the Boll Weevil*, ed. R. L. Ridgway, E. P. Loyd and W. H. Cross, 437-479. USDA.

Brown, L. G., J. W. Jones, J. D. Hesketh, J. D. Hartsog, F. D. Whisler and F. A. Harris. 1985. COTCROP: Computer simulation of cotton growth and yield. Information Bulletin No. 69. Mississippi Agricultural and Forestry Experiment Station.

McDaniel, S. G. 1976. Rate, temperature, *Heliothis* spp. and developmental stage effect on methyl parathion efficacy in cotton. M.S. thesis, Mississippi State University, Mississippi State.

Index